Scientific Literacy and Environmental Policy

Scientific Literacy and Environmental Policy

The Missing Prerequisite for Sound Decision Making

Dorothy J. Howell

Q

QUORUM BOOKS

New York
Westport, Connecticut
London

Library of Congress Cataloging-in-Publication Data

Howell, Dorothy J.
Scientific literacy and environmental policy : the missing prerequisite for sound decision making / Dorothy J. Howell.
p. cm.
Includes bibliographical references and index.
ISBN 0-89930-616-0 (alk. paper)
1. Environmental policy—United States—Decision making.
2. Environmental policy—United States—Citizen participation.
3. Science—Study and teaching—United States. 4. New products—United States. I. Title
HC110.E5H688 1992
363.7'058'0973—dc20 91-36028

British Library Cataloguing in Publication Data is available.

Library of Congress Catalog Card Number: 91-36028
ISBN: 0-89930-616-0

First published in 1992

Quorum Books, One Madison Avenue, New York, NY 10010
An imprint of Greenwood Publishing Group, Inc.

Printed in the United States of America

The paper used in this book complies with the Permanent Paper Standard issued by the National Information Standards Organization (Z39.48-1984).

10 9 8 7 6 5 4 3 2 1

TO MY TEACHERS, COLLEAGUES, AND FRIENDS OVER THE YEARS

A constant source of inspiration and encouragement

Contents

Acknowledgments

No book ever emerges exclusively from the mind of its named author. This one has evolved from a variety of sources actively assimilated into the author's professional outlook. It is only right to recognize those who have contributed most positively to whatever advances may be prompted toward scientific literacy on behalf of environmental policy. Additional impediments to enhanced communication I must accept as mine alone.

Blessed with a family who imposed no limits on aspirations or intellectual development, I have thrived in expanding horizons of literacy. Among those for whom the bumper sticker should read, "So little time—so many books," I have a mind constantly nurtured by books and libraries. Librarians have been good to me, starting with Edith Nettleton of Guilford Public Library in Connecticut. Today I enjoy the bounty of Vermont Law School's superb library. The staff, represented by Diane Frake, provide skilled services complemented with a congenial spirit.

Good, bad, or indifferent—I owe something to every teacher encountered all the way through a master's and part of a Ph.D. in the biological sciences to the J.D. Sometimes individual lessons have revealed a need for improvement, but most teachers have truly been mentors. Together those in Guilford and at Baltimore's Western High School opened the academic world, the latter through its advanced college preparatory program. The names Jillson, Potter, Kul, Griffin, and Troy stand out. But my professional education actually began with Maryland's Goucher College, where I encountered mentors in Professor Helen Habermann and the late Helen Funk, professor of microbiology. Goucher actively encouraged graduate work in the biological sciences.

Applied microbial ecology at Chicago's Metropolitan Sanitary District drew me out of academia to discover the real world, including environmental protection and pollution control, in part through active participation in public response to water pollution matters. Not the least of the lessons was the inability of the legal community and the professionals of science and technology to communicate with each other or with the public. I was eyewitness to the inevitable frustrations. These lessons led me to cross professional lines to enter environmental law, with the financial support of the Sanitary District. The people who stand representative of the best of that era are the late Stanley Whitebloom, William and Alice Schmeelk (and their family), and Samuel Dennison. John Marshall Law School furthered the ambition to foster communication across disciplinary lines as an environmental professional and is best remembered through John Sheid and Samuel Lawton, the latter a "founding" member of Illinois' pioneering Pollution Control Board.

After a brief stint as an assistant in the environmental program of the Illinois Attorney General, I proceeded to a career as corporate environmental counsel, first with Borg-Warner Corporation and then with Chemical Waste Management. From those vantages, I experienced the panorama of evolving pollution control law from the mid-seventies through the mid-eighties. It was not unusual for the same "pollutant" to be under scrutiny as a risk to environmental health at three levels—ecological, workplace environments, and the human organism—in programs administered by the U.S. Environmental Protection Agency and its local counterparts, the Occupational Safety and Health Administration and the Food and Drug Administration.

Professional integrity and dedication to a sagacious balance between environmental stewardship and marketplace reality were exemplified by colleagues such as William Stiens, William Reagan, Robert Gilligan, Robert Miller, Dixie Laswell, Johnine Brown, and Jodie Bernstein. Further inspiration was provided by Howard Chin, David Flannery, Joseph Hadley, Donald Cherry and, at a greater distance, John Cairns, Jr., who writes on the parallel tracks of ecological reality and pollution control policy in the hope of the convergence promoted here.

It was in corporate practice that utter frustration with the elevation of political resolutions imposed by government at the expense of science and technology, not to mention applicable law and policy, finally took their toll. In the second Decade of the Environment, the increasing demand for communication among interested professionals and between them and the public remained unmet.

Personal advancement of a more effective merger between science and law was activated through a Bell and Howell Postdoctoral Research Fellowship at Boston University's Center for Technology and Policy, where initial drafting of this book commenced. That there is hope for scientific literacy to advance interdisciplinary communication in aid of sound environmental decisions is reflected in the existence of the Center and the fellowship. That the road ahead will continue to be rocky is reflected in the absence of a second offering of fellowships and the resorption of the Center into a more academically "respectable" department.

Since 1988, I have been privileged to be a part of the Environmental Law Center (ELC) at Vermont Law School, with a year "visiting" at Franklin Pierce Law Center, where law and science meet in a more traditional sense. Scientific literacy can only be advanced in such programs. That the two schools have recently instituted an exchange program may be a sign of smoother patches of highway ahead.

One who reflects the best of ELC is Richard Brooks, a founder and former director. It is he who is most immediately responsible for active reflection on the interdisciplinary advancement of environmental stewardship. It is fitting that Director Brooks has been succeeded by two—one a lawyer, the other with a doctorate in public policy. In less than a year Jonathan Lash and Rick Melberth are already welcome additions to the rank of mentor.

Finally, I am grateful for the sharp blade of Eric Valentine's literary scalpel for incisive intervention during the final stages of drafting.

Introduction

If there is an intellectual fate worse than death it is illiteracy. For environmental professionals and their counterparts at other national intersections where science and technology are meeting the politicolegal system, a comparable fate is scientific illiteracy. In a very real sense, scientific illiteracy has disenfranchised society in the United States in decisions ranging from personal health reached by the Food and Drug Administration to environmental quality as effected by the U.S. Environmental Protection Agency. That there is little hope for sound policy formulation in these programs until nationwide scientific literacy is actively practiced is the thesis of *Scientific Literacy and Environmental Policy*.

Education and personal experience with no little frustration have led to this thesis and its exploration here. Although this book draws heavily on a variety of professional commentary, the author's own experiences have provided vivid illustrations of the quandaries posed by widespread scientific illiteracy. In a sense, this effort is a scholarly annotation of that experience: There is nothing here without some parallel in the author's activities as a scientist, environmental counsel, or interested citizen. Her personal vantage has gained perspective through tenures in the corporate world, complemented by those in government, academia and briefly, in association with groups holding themselves out as representatives of that special interest identified as "the public."

Personal experience lends immediacy to the commentary, while the commentary adds legitimacy to the experience. Together, experience and contemporary commentary—spanning the intellectual range from scholarly dissertation through apologia to popular diatribe—suggest a rare convergence of thought: In the rush to realize the promises of the scientific revolution, this country has effectively disenfranchised an electorate suffering the ravages of scientific illiteracy.

With astonishing acceleration scientific advances are entering the marketplace without much opportunity for public reflection. Perceived hazards may subsequently come before our institutions of policy formulation, only to be resolved in the absence of informed participation by society at large. In the marketplace and throughout the politicolegal system, policy evolves through a balancing among vested interests. Underlying realities of science and technology are propounded with little regard for their effective communication across disciplinary lines or to the lay public. The result is frequently beneficial to one particularly articulate sector at the expense of others, including the larger society. In the process science and technology themselves become garbled, if not lost entirely.

The quandaries before us are characteristic of national decisionmaking associated with two "expert" administrative agencies: the Food and Drug Administration (FDA) and the U.S. Environmental Protection Agency (EPA). Pharmaceutical innovation and environmental protection, the latter with specific reference to products of molecular recombinant DNA (biotechnology), are representative of policy decisions at the intersection where science, technology, law, and policy are meeting. As such, these decisions go beyond revelation to provide impetus for scientific literacy in advancement of sound policy formulation in response to future scientific innovations.

In Sheila Jasanoff's 1990 book, *The Fifth Branch: Science Advisers as Policymakers*, the intersection where science encounters the politicolegal system is explored in light of the "troubled marriage" between law and science. Jasanoff recounts interprofessional exchanges in a variety of our political institutions, notably including FDA and EPA. She proposes improvement to facilitate smooth relationships while avoiding the pitfalls of past experience. In their book of the same year, Marc K. Landy, Marc J. Roberts, and Stephen R. Thomas explore much the same territory from the political perspective through experiences at EPA. In *The Environmental Protection Agency: Asking the Wrong Questions*, they too propose improvements.

These four observers and their countless colleagues perceive from unique perspectives the needs demanded by the intersection. All the observers offer valuable insights, limited, however, by the respective vantage of each observer. In the end, although each commentary is insightful, professional perspective is one limitation among many impeding progress toward needed reforms. Many impediments are individual, but others are institutional. As such they may ultimately prove intractable. Meetings at the intersection may well produce little more than gridlock.

A common perception among those familiar with the practices of FDA and EPA is that perceptions override realities as policy takes on form and substance. In large part the difficulty can be attributed to unrealistic and inconsistent demands from constituencies who do not quite trust the agency and certainly distrust their own counterparts.

Much of the clash at the intersection is probably unavoidable. It is the product of our carefully counterbalanced institutions—societal, economic, and political. We have little desire to retreat from either capitalism or our constitutional democratic republic. It may be that contemporary bureaucracy is the only tool that comes close to functioning in face of basic human nature. But some improvement would follow a better appreciation of underlying science and technology and for scientists and their methods. If nothing else, appreciation opens the way to more effective communication.

The literature chronicling events at the intersection is mammoth and growing, with no end presently in sight. The literature includes, but seldom incorporates, the parallel literature of scientific illiteracy in this country, one element of the more pervasive pall of societal illiteracy in general.

What is missing throughout, however, is an exploration that itself clearly crosses disciplinary boundaries to offer an integrated, interdisciplinary response reflecting the many facets and doing true justice to the complexities of major national crises. Perhaps, after all, the clearest perceptions will reside with the interested citizen with his or her feet in more than one of the forums and loyalties extending beyond institutional constraints or personal vantage point. *Scientific Literacy and Environmental Policy* endeavors to provide just that perspective in language neither condescending nor loaded with jargon.

Acknowledging effective disenfranchisement is the first step toward reenfranchisement. The requisite second step is practical scientific literacy throughout society. In support of this thesis, scientific literacy is subject to an ambitious definition: On the one hand, it assumes the abilities to read with comprehension news items on science and to apply scientific information to personal decisions on emerging policy based in science. On the other, it instills the capacity for meaningful participation in policy formulation through recognition of relevant issues with an appreciation of the underlying science and technology, including a realistic view of their limitations.

We are far from any such achievement.

Here we will explore the need for scientific literacy in environmental policy formulation, itself broadly defined to include programs of FDA as well as EPA. First, the two relevant cultures, the U.S. political system and the sociologies of science, are established and their contemporary merger examined in historical perspective. Thereupon, policy formulation in face of public controversies associated with pharmaceutical innovation is reviewed, followed by review of environmental policy formulation that arises from public controversy prompted by biotechnology. These reviews demonstrate the disenfranchisement of the major part of the electorate.

The scope of disenfranchisement is explored further through the writings of observers within and outside relevant institutions. The culprit, scientific illiteracy, is expressly recognized or at least implicit throughout the commentary. Hope for reenfranchisement is actually found in existing institutions, notably those of formal education and the news media, but ingrained traditions are also recognized as major impediments to establishing scientific literacy in aid of needed reform. Finally, an integrated program for advancing scientific literacy is gleaned from the wholly disjointed efforts in which the occasional inspired individual and a few enlightened organizations are engaged.

This book is by an environmental professional for all interested professionals, but it is intended also for any individual actively concerned about the places of innovative science and technology among our institutions. The sources have been selected accordingly. The reader is encouraged to explore the annotations and thereby expand his or her own intellectual horizons.

Because the influence of scientific illiteracy demands focus as much on the perceptions as on the realities at the intersection, certain sources have been selected for their relevance to coping with the superimposition of perception on realities of our technological and political institutions. Most notable are *Science* and the *Wall Street Journal.* Both are available to the public and, for the most part, accessible, that is, readable by those uninitiated in the participating disciplines. Both represent vested interests, albeit broader than most.

The avowed mission is to contribute to reenfranchisement, not to compound the disenfranchisement effected through scientific illiteracy in league with private agendas, remote science, unresponsive markets, obtuse government, and pervasive distrust. Only with scientific literacy can perceptions come closer to reality and public demands approach real needs, now and as projected into that future promised by gathering momentum in the scientific revolution. All the participants have both a stake and a role here; each must be prepared to rise above personal impedimenta to fulfill the role.

PART I

THE SETTING

CHAPTER 1

Prologue: Regulation in General

SYNOPSIS OF FEDERALISM TODAY

"Congress just doesn't seem to work these days."[1]

Basic to our democratic republic is the active participation of the electorate in governing. Public policy is the responsibility of the people it is intended to serve. Participation is expected (if not invited or actually demanded) in all areas of formal policymaking, from legislation through adjudication of its provisions in court. But no one can deny the recent emergence of novel problems inhibiting meaningful involvement by the public at large. These fundamental problems are exacerbated whenever a national policy is evolving at the intersection where law and politics meet science and technology.

It is the intent of this book to analyze these problems primarily in association with two major technologies subject to U.S. regulation: pharmaceutical innovation and biotechnology. Both possess elements of novelty. Evolving regulatory approaches must rise to the resultant challenges if they are to serve society. Although policy formulation affects these innovations, that process takes on still greater import in pointing the way to regulatory response to the inevitable innovations in our future.

Questions come immediately to mind: What is the nature of public participation? How effective is it from the vantage points of the individual participant, of the policymakers, *and* of society? How can public participation be improved? Are there lessons here to be applied to policy formulation in future arenas arising from similar, equally unprecedented advances?

No consideration of the evolution of policy at its intersection with science and technology can proceed in a vacuum. Without some recognition of the unique political setting of contemporary policy formulation, it is impossible to sort out the troubles. Without some comprehension of the historical context of modern science and technology, most members of the public are simultaneously dazzled and dismayed, depending on individual perceptions of reported advances.

Without acknowledgment of the roots of today's American political system, the unique features of the contemporary scene remain elusive. Worse, the system presently conveys the impression of standing well beyond the touch of individual members of the electorate.

Today, the individual seems distanced from both political process and technological advances. Circumstances tend to place the political forums seemingly out of reach. The layman holds science at a safe distance. As a result the individual is not merely isolated, but also estranged from the institutions of both systems.

But when some discipline of science or technology is brought home to the individual by way of an issue that can be resolved only through the political process, individuals promptly find themselves thrust into controversies demanding instant answers, without reference to the capacity of either system to provide them. The matter of capacity notwithstanding, some answer will be forthcoming.

Thus, members of the electorate are treated to the spectacle of the politician trying to regulate without a sound technical basis; or the judge deciding a major case, where technical experts have accomplished little more than confirm the uncertainty; or the scientist who cannot give a simple "yes" or a simple "no" but must endlessly qualify and equivocate on a matter of crucial public interest. Despite isolation and likely estrangement, the individual may well be required to make a politically significant decision on the issue. That decision may take the form of a vote, a verdict at trial, or a position in a local controversy with national ramifications.

Before addressing such untenable situations, it is wise to examine both the political and the scientific setting, each in the context of its phenomenal recent history.

Federalism as established by the U.S. Constitution

> Majority rule balanced by minority rights gives "rise to concern for citizen education for broadening the basis of public understanding of policy alternatives."[2]

If we the people are doing our several jobs as citizens, every high school graduate in America is familiar with the basics of federalism, the formal relationship by which power is distributed between the federal government and the several state governments. In fact, the concept of federalism in this country is incomplete in the absence of distributions of power between the people and their governments and within the U.S. government itself.

The Constitution authorizes the federal government to pass only those classes of laws expressly enumerated, all national in scope.[3] For our purposes, the most notable among these is the power to regulate commerce among the several states. Ultimately, "[t]he powers not delegated to the United States by the Constitution nor prohibited by it to the States, are reserved to the States respectively, or to the people."[4]

In an intricate system of checks and balances on the powers distributed within the U.S. government, the three branches are established.[5] The two houses of Congress initiate lawmaking through the introduction of a bill, the specific contents of which must be approved by both houses before proceeding to the President. Once a bill is signed into law, it is the Executive which enforces it, often through regulations promulgated by various administrative agencies of the Executive, known collectively as the bureaucracy.

In addition to the body of decisions arising from legal theories based originally in English common law and equity, the judiciary is responsible for interpretation of the language of statutes and regulations and for determining the propriety of their execution. These "codified" laws, upon challenge through litigation, must pass constitutional muster as written and in their implementation, both generally and in specific circumstances. Even when a statute or regulation does so, both it and the process and substance of its execution remain subject to interpretation. Judges can in effect revise or nullify either a given provision of law itself or the execution of it.

In considering the American ways of policy formulation, it is essential to remember two specific substantive provisions of the Constitution: the Supremacy Clause[6] and the Commerce Clause.[7] The existence—and the validity—of all the law to be discussed here depends entirely on those two clauses. Each law must represent a legitimate regulation of commerce in order to take priority over state law or over the rights and privileges of the people. Thus, all those who would participate in the formulation of policy must never lose sight of those constitutional sources. To some the philosophical connection between a regulatory program and this primary source can seem tenuous at best.

Public participation in political processes is itself a guarantee of the Constitution as amended through the Bill of Rights.[8] The electorate has the right and the duty to participate in government. Nowhere in the Constitution, however, is there established a concomitant duty to be informed on the issues. Moreover, there is no *constitutional* authority to ensure the education of the public. No doubt the Founding Fathers had no reason to believe public education and the press would ever fail in their respective contributions to this crucial need. At one point in history, those lapses may have been of little import. That has changed today, at least in part because government has changed within its constitutional framework.

THE REGULATORY PROCESS IN THE EIGHTIES AND NINETIES

> "Given the economic importance and the shrillness of the political debate about the burdens of regulation, it is surprising how little serious research exists on both the direct and indirect effects of regulation on productivity growth."[9]

Never quite as simple as might be inferred from the exquisitely succinct language of the Constitution, today's politics is characterized by confounding factors unique to the latter half of this century.

Though identity with a political party has always been an important part of American life, today a case can be made for ideologies and political ideas as replacements for party as the focal point of individual loyalties.[10] Today's activists are striving to see their principles reflected in public policy. The resulting image becomes a politics of ideas rather than of material interests. It is worth noting in the context of current events that an ideology is deemed far more demanding on its followers than any political party.[11]

The emergence of a politics of ideas reflects profound changes in the American electorate; the impact on American policy formulation can be expected to be equally profound. Attributed to such factors as an American people who are more affluent, more mobile, and better educated, the alteration in philosophy is also held responsible for an increasing "nationalization" of politics, that is, greater emphasis on the U.S. government itself as the source of public policy. Moreover, face-to-face communication and relationships have been replaced with impersonal communication, primarily through television.[12]

In effect, this emphasis on ideas has led to the decline of state and local political party organizations. Their replacements are organizations composed of individuals who are well educated, highly principled, and activist, who decry compromise and acceptance as decadent and immoral. Their memberships are comprised of those who are active on a day-to-day basis in politics held to be centrally important to the survival of democracy. This political class is not drawn exclusively from the elected officials, top bureaucrats, and judges, all of whom were contemplated at least in part by the Constitution. Along with the more familiar political parties and their organizers, this new class expands to include reporters, editors, lobbyists, and interest-group leaders.[13]

A potentially greater concern follows from the observation that political systems based on ideology are notoriously brittle. When inflexibility becomes nationalized, the losers in the array of contests may no longer sit back and await the next round.[14] One can only wonder whether the public at large is truly represented.

In the Congress, redistricting and a larger electorate have altered elections and politics. But today's new political technology, affecting communication, financing, and organization, has opened the doors to insurgent candidates. From the mid-sixties the more genuinely national political system has resulted in stronger individual legislators. But Congress is deemed weaker, less than the sum of its parts.[15]

During the same interval the bureaucracy (ordinarily identified as the body of administrative agencies within the executive branch) has become more independent and more powerful. Through the bureaucracy, the national government has assumed responsibility for a wide range of new programs, notably including environmental protection. Because the bureaucracy has not enlarged so much as it has strengthened through delegation to state and local agencies, the informal, de facto controls once practiced by state and local politicians have been supplanted. Once again power has become nationalized, this time in the Executive.[16]

Administrative agencies act through regulations promulgated under the aegis of statutes. Regulation, like legislation, can be economic or social in concept.[17] The effects of regulation are measured in costs, benefits, and effect on productivity. Pharmaceutical innovation and, especially, environmental protection comprise a major component of today's social regulation.[18]

Costs are a major and particularly difficult component of social regulation, especially in the face of multiple unknowns. There are three components of regulatory costs: the administrative expenses incurred by the responsible agency itself, the direct costs of compliance, and the indirect costs attributed to inducements associated with a regulation, for example, the inducement not to build at an otherwise low-cost location.[19] Upon no more than cursory study, two observations become abundantly clear: The bureaucracy is itself big "business," and the associated costs are subject to considerable controversy.

It is worth noting that the reform legislation of the early seventies, notably including the environmental mandates, were more specific than many of the predecessor programs. The intent is said to be an effort to render this legislation less subject to judicial interpretation.[20]

Notwithstanding the proliferation of regulatory programs dedicated to the protection of the environment and of human health and safety, our society has always been ambivalent about such government intervention. While espousing a freely competitive market, we remain eager for an array of governmental assistance efforts. Through the sixties and seventies the upper hand was with proponents of intervention. The result of this ideological fervor, manifested in numerous sectors, was a veritable explosion of regulation.[21]

Despite the fact that the political phenomenon of social regulation first took form in the early part of this century, nearly all the major programs passed in the sixties and seventies represent social regulation. Salient among these are the Food and Drug Amendments of 1962, the Clean Air Act of

1970, and the Toxic Substances Control Act of 1976. Altogether some forty major legislative items of social regulation were passed over sixteen years.[22] As a result of this contemporary proliferation of social regulation, we presently are laboring in a complex and loose-jointed system of issue networks rather than the earlier "iron triangle" of bureaucrats, interest groups, and congressional committees.

At least on the surface, these issue networks represent a system more open to public participation. The triangle has been opened to include lawyers, journalists, advisory panels, state officials (elected and otherwise), officials of nonprofit organizations, professors, and experts. Although direct material interests are still represented, they are often secondary to intellectual, emotional, or ideological stances. But, rather than some construct of the public will, the common denominator here is a specialized knowledge about policy alternatives.[23]

This situation may prove small comfort to the interested individual member of the electorate. New networks may open the door to a wider variety of views than was previously the case, and better academic training may back participants than was the case with politicians of an earlier time; but these new policy debates are frequently abstruse and obscured in statistical analysis, academic writings, the offerings of think tanks, and dreary congressional hearings. While meaningful to the experts, they are all but impenetrable to most of the electorate. Hence, the capacity of the electorate to follow and to judge policy questions is reduced.[24] In fact, the "noncombatant" public at large may have been driven out, leaving the process to the zealots of the extremes and further isolating the bureaucracy from popular control.[25] Among these zealots are the public interest groups, a phenomenon of the eighties, but hardly new to politics.

It has been said that James Madison, writing as "Publius," had reference to interest groups when he spoke of "factions" who would seize control of government. Clearly, then, interest groups have always assumed a fundamental role in American politics. But they have not necessarily been popular or even trusted. In fact, there has been skepticism about their motives and outright hostility to their influence.[26]

Today's version of these factions found its genesis with Ralph Nader, who

> appears to have been the first political entrepreneur to fully appreciate the central institutional lessons taught by the civil rights movement, which were that leaders in Washington could be subjected to direct pressure through the use of television and the mobilization of activists, and that where these techniques failed, appeal to the courts might succeed.[27]

A major feature of today's political system, with special reference to the bureaucracy, is inefficiency. Two sources have been cited: Initially, statutory inhibitions on agency efforts to balance benefits against costs, in the face of rigid command and control approaches, and, second, the lack of oversight. Although they must work within the constraints of their statutory powers and a broad social consensus, bureaucrats have in effect been issued a blank check in terms of the private costs of compliance. Thus, decision makers are not forced to choose among competing programs within limited budgets. This unlimited line of credit on the private purse is unique to today's regulatory agenda. Moreover, congressional oversight is weak and sporadic. It, too, proceeds in the absence of priorities.[28]

Finally, because of the lack of any mechanism to force elected decision makers to effect a balance among regulatory objectives,support for reopening a particular statute lends itself to characterization as an attempt to weaken that statute. The least suggestion of allowing agencies greater freedom to make trade-offs in their decisions can therefore be labelled as promotion of, for example, dirty air or unsafe workplaces. If, instead, elected officials were able to convey the reality that the pursuit of cleaner air must be balanced against safer disposal of toxic wastes, as is already accomplished in choices between defense and social programs, the intractable political issues involved in amending restrictive statutes would be more amenable to resolution.[29]

JUDICIAL PROCESS IN THE EIGHTIES AND NINETIES

> "[T]here has been a rush to the courtroom by political litigants hoping to win before the bar what they have been denied in the give-and-take of the political process."[30]

The characterization of the Congress as "weak" notwithstanding, from 1968 through 1982 over a hundred new legal claims were established by new statutes. Some of these claims were entirely novel to the judicial system. Although it is the passage of significant bills that commits the government to new responsibilities, it is the courts that will eventually decide important cases and establish new law in concrete, enforceable terms. The courts are called upon to resolve the ambiguities and to decide what a statute really means. Because the elected officials are thus insulated from painful choices, Congress goes along with the resultant increase in litigation,[31] subject, of course, to the constraints imposed by legislative history and precedent.

Over the past twenty-five years the courts have moved closer to being explicitly political.[32] Attorneys and judges increasingly act as policy makers and arbiters. Obviously, the courts have thus become more open to political questions than ever before. The tradition of leaving issues to the legislature and the political process is eroding. Congress, itself, takes advantage of the trend in ways both explicit and subtle. In the latter case legislative language is ambiguous; in the former decisions are frankly delegated to the administrative agencies.[33]

Hence the rise of public interest law firms—first those leaning to the left, more recently those leaning to the right. It can come as no surprise that attorneys have followed the rest into the practice of ideological law. This phenomenon serves to swell the workload on the courts still further and increases the number of attorneys.

Now the groups constituting the U.S. Supreme Court's prime audience are the intellectuals who comment on legal issues, other attorneys and judges and articulate politicians. The Supreme Court hears the public at large only through the filter of that highly educated composite.[34]

In fact, litigation as a political tactic has been encouraged. Policy-conscious litigation is subsidized through reimbursement of private attorneys as "private attorneys-general." It is the opposing party who does the reimbursing.[35]

But a major problem has been recognized: When activist law firms are mentioned, "public interest" belongs in quotes. Who knows where the public interest actually lies in the issues so hotly debated? The term alone is virtually impossible to define. One of the early "public interest" lawyers in Washington was candid about the way he defined the term: "a public interest case is anything that turns me on, that serves a cause I believe in."[36] While attorneys who fight battles in the "public interest" are characterized as sincere, the question remains: Is the courtroom the proper forum for debate and resolution of politically significant questions?[37] There have been fashions in litigation. While "old-fashioned" cases are still being heard, the new are being superimposed (and have their effect) on the old. According to William N. Lunch, one result in the U.S. Supreme Court is the phenomenon of far more plurality decisions, often with numerous concurring and dissenting opinions.[38]

REALITIES OF THE CONTEMPORARY REGULATORY PROCESS

> "The mandates of regulatory agencies do not allow them any rest if any possibility exists of their being considered derelict in a responsibility."[39]

The Milieu

> "Each generation tends to bend the bow too vigorously in a given direction in order to counteract the excesses of its predecessors in the opposite one."[40]

The preceding review has provided some insights into the condition of the contemporary political milieu. While unmistakably based in the precepts of the Constitution, today's political process reveals a peculiar character all its own, which character must be taken into account whenever addressing any specific substantive matter. From the unique character of modern politics spring the realities that must be accepted throughout the examination of public participation in the regulatory process.

Today the Congress, while said to be composed of stronger individual legislators, is itself deemed weaker than ever before. In contrast, the executive branch, in the form of the bureaucracy, is extraordinarily strong from the historical perspective. Consequently, the courts are increasingly superimposing political decisions on their more traditional adjudication of cases, historically quite limited in scope. Ultimately, the public is increasingly isolated from political processes and even estranged from government, despite the fact that so-called public interest groups are flourishing as never before.

Here the implications of this contemporary political situation will be considered in greater detail, in preparation for the corresponding exploration of the contemporaneous revolution in science and technology.

A useful starting point is an examination of the pattern followed by most social regulation.[41]

First, the basic reform is enacted. Immediately, it becomes necessary to take note of the argument that it is arithmetically impossible for the vast majority to receive special privileges. On any given issue the stake of each member of the public as a taxpayer and consumer (presumably constituting the "vast majority" on that issue) is small in comparison to the stake of the competing groups with their respective vested, special interests. Therefore, the transcending reality is that some concentrated private interest usually will prevail over the diffused general interest.

Whomever its supporters may actually represent, the reform is effected through a new administrative agency, established by mandate of the reforming statute. Initially, members of the agency's staff are embued with devotion to the reform movement, and they proceed with tremendous enthusiasm. Their administrators are in deep sympathy with the movement. But, ironically, the pioneering reformers have ordinarily been discredited in the process of effecting the change and are unable to hold public office.[42]

As time passes many of the pioneers disappear from the bureaucratic scene. Those who remain and retain their vigor come to be attacked, broken politically, and discredited. Then, over time, more subtle opposition and even disgruntled proponents come to the fore. Any diehard pioneers commence to see the realities of the situation, grow tired, or merely lapse into dull routine.[43]

Moreover, public interest slackens: The objective has been won, and the effort now enters the stage of incomprehensible technical questions. Meanwhile, of course, private interests continue their pressure on the agency.[44]

Standards of Public Performance and Barriers to Meeting Them

> "At the simplest level, the proliferation and overlap of regulatory agencies are routinely blamed for some of the regulatory horrors that are often produced to show what a mess the system is in."[45]

These realities notwithstanding, the general level of bureaucratic conduct remains relatively high. The majority of bureaucrats are not only honest but also motivated to perform effectively. They work hard and are paid less than they would be in private industry. Ordinarily the public is unaware of them or actually ignores them.[46]

But those basic difficulties, recited so eloquently by Senator Douglas forty years ago, are compounded in highly technical areas by the additional uncertainties and difficulties of balancing costs with benefits. Furthermore, there are relatively few incentives to encourage improvement of existing efforts of such estimates in either the public or the private sector.[47]

In a sense, then, it is the nature of the political beast both to establish the standards of excellence and simultaneously to doom regulatory programs to a languishing demise. The milieu itself is conducive to both accomplishments.[48]

The low esteem in which the public holds Congress is one untoward result of the circumstances arising from the contemporary amalgamation of public isolation and political impotence with the gradual draining of spirit from agency personnel:

> Congress remains . . . a mirror, but it is now a shattered one, producing kaleidoscopic images, individually attractive but collectively disconcerting and fragmented. Perhaps it is only reflecting the nation, but if the public is unhappy with Congress and the members themselves are frustrated, it seems fair to quote Ralph Waldo Emerson . . . "The sower may mistake and sow his peas crookedly; the peas make no mistake but come up and show his line."[49]

Returning again to Senator Douglas, we discover that the undesirable occupational characteristics of government employment are hardly unique to our own era.[50] The bureaucracy has always heightened the desire for security while dulling the initiative. The pace is slow and marked with much malingering: Dismissal for cause is no easy task. The tendency is to play it safe and to remain suspicious of new methods or of any challenge to old procedures. In the effort to avoid mistakes innovation is shunned and new responsibilities are refused. Centralization is encouraged and decisions dehumanized with undue absorption in paperwork. Decisions gain distance from affected interests and persons. It can be no cause for wonder when agency work tends to deteriorate with time.

"Administrativitis"

> "When elected political leaders feel they can act irresponsibly because they know that unelected judges will be responsible, should we be surprised that the cynicism about democratic institutions grows?"[51]

What is new in contemporary America is the high-minded legislation initiated in the sixties and promising to proceed into the twenty-first century. This novel effort has left it to the bureaucracy to transform goals into workable rules and regulations. The inevitable result has been an explosion of litigation. The courts are entering the arena to decide which provisions are to be enforced, what exceptions are warranted, and so forth. Final decisions, then, come out of litigation rather than directly proceeding from statutory language or regulatory mandate. But, in fact, the hallmark of these codified requirements is ambiguity.[52]

Ambiguity reflects political compromise. The inevitable result of compromise is, in turn, a series of transformations: The specific becomes general, the precise turns vague, and clarity is rendered muddy. The buck of resolution is initially passed to the bureaucracy.[53]

Nevertheless, regulatory effort has yet to be subjected to any coordination. It shifts with shifting political and ideological trends. Instead of reacting to market failures, regulations have served to redistribute societal resources toward sectors who have successfully petitioned their government,[54] in time-honored constitutional tradition!

But, as Senator Douglas predicted:

> Gradually . . . the once virile agency loses its crusading spirit and becomes both lethargic and indifferent to the public interest. This may well be a general law of life. It is certainly the history of most governmental regulatory bodies. I call this degenerative disease "administrativitis." The only antidote is the continued injection of new and vigorous personalities, active work by the lonely soldiers of the common good, and, *most important of all, the constant and informed interest of the general public.*[55]

There are two levels of response to the regulatory morass in which we find ourselves: the official one and that of the public at large. The sources of official response would be expected to be the Congress and the judiciary. As has already been noted, however, Congress is more the original source of the problem than a likely reformer. In any event, as litigation is the initial route to any reform, it is logical to look first to the courts for relief. But, in spite of—or, perhaps, as a result of—the litigation explosion, the courts, too, appear to be more a part of contemporary problems than a correcting factor. They are themselves subject to severe criticism by observers as a result, at least in part, of the trends already noted here.

But there is a larger criticism. It is not so much the allegation that the courts have been forcing on a resistant public grossly unpopular policies, however true that allegation. The more significant charge is one of short-circuiting the democratic process. Once such moves resulted in legislative change in the states, with all the attendant publicity and exchanges of positions among all interested sectors and resultant compromises. Although politics was consistently slow, frustrating, and even painful, that is the way a democracy goes about making up its mind. And legitimacy was much better served.[56]

In fact, it has recently been argued, law has become so complicated that there is always a technical flaw in any decision, administrative or judicial, or some distant case precedent, either of which will support *any* judicial decision. The result is a nation of men, not laws, and political rather than legal decisions. Though a reasonable resolution may ultimately be reached among policy makers, the essential predictability of the system is lost.[57]

What, then, is the public response? Is it more likely to lead to reform than the responses of its representatives? That is unlikely, according to at least one member of Congress. This lawmaker distributes newsletters to his constituency. At one time issues were discussed, but the constituents did not read the newsletter. The format was altered accordingly. No longer is it dedicated to content; instead, it is no more than a source of name-identification. To paraphrase that lawmaker, distributing the newsletter is akin to selling a bar of soap.[58]

From the milieu under discussion the suggestion emerges that the public at large is being supplanted in the political process by the special interest groups.[59] While claiming issues of public policy should be determined on the merits rather than reflect the accidental by-product of some struggle for personal and party advantage, these groups, like the government, are distanced from the public at large. In fact, they themselves are less and less likely to reflect any geographic or economic diversity throughout the nation.[60]

Given the nature of special interest groups as one focus of power in the bureaucracy, at the expense of Congress and with the increasing political intrusions of the courts, it is most disturbing to note that, in instances when reporters or television can be interested, a small special interest group can generate as much press for its limited view as a bureaucracy representing millions of constituents. Thus, the "new politics of information" can give rise to significant distortions regarding relative strengths among vested interests on a nationwide basis. With their own respective interests in sales, those media representatives will be attracted to the outlandish, whatever the form—politicians, charges against bureaucrats, proposals for new policies. It stands to reason that those skilled in the emergent politics of information know and exploit these realities.[61]

Can it really be that both the public and private sectors have abdicated responsibility and allowed it to devolve to special interests and the fantasies of the press?

IMPLICATIONS

> "James Madison purposely built inefficiencies into the system to protect against rash decisions. But in the view of many scholars and lawmakers

themselves, the problem today isn't simply inefficiency; it is an inability to deal with the nation's problems."[62]

The major lesson of the political realities collectively dubbed "administrativitis" is that the supervision and control prerequisite to effective regulation are lacking. Although individual elements are supervised, no branch of government systematically examines either the broad priorities or the overall effect of social regulation. No one has ever undertaken the daunting task of determining the effect in aggregate in terms of either benefits or costs. The diverse statutes are managed by their separate agencies, each promulgating dozens of rules every year.[63]

An overseer is not to be found in any of the three branches of government. While Congress controls appointments, authorizations, and appropriations and has the power to review specific rules, it is preoccupied with new business and does not use its powers effectively. Moreover, there is no institutional mechanism of needed oversight. The courts can move only on individual decisions actually before them. Surprisingly enough, the Executive itself, home of the bureaucracy, has taken some steps, but its attempts lack the coordinated effort needed to establish priorities.[64]

The implications are both political and economic. On the political front, decisions of fundamental importance, ultimately establishing which groups in society will most benefit, are either left to the unelected officials of agencies or are simply not decided at all. On the economic front, inefficiency follows because there is neither any incentive to balance objectives with associated costs nor any requirement that private dollars go first to programs producing the greatest benefits.[65]

Can there be any wonder at the declaration that Congress no longer works? Or that at least one senator has expressed a sense that the whole system is breaking down?[66]

While exceptions do exist, Congress is perceived, from within as well as without, as doing more and more work to produce less and less. Faced with the array of impediments to effective lawmaking, the membership turns advocate at the expense of statesmanship.[67] More telling than such observations is the lack of any recommendation for correcting the situation. Increasingly, crucial decisions are made under pressure and with a disturbing combination of influence, favoritism, improper deals, and outright corruption.[68]

At the same time Congress has come to be deemed impotent, the wild expansion of litigation and the consequent expansion of the influence of the legal profession at large has been viewed with alarm.[69] First and foremost, that alarm proceeds from the knowledge that the legal profession is not elected to the exalted status it enjoys. Where the right to govern depends on the consent of those governed, the courts should not do so. If the public does make fundamental political or social choices only to have them routinely upset or overturned completely, citizens are deprived of any sense of a collective role in those choices."[70]

More insidious is the tendency among legislators and agency management consciously to enter unwise decisions to satisfy passionate minorities, secure in the knowledge that the decision will undergo review. At least two more results follow:

> When elected political leaders feel they can act irresponsibly because they know that unelected judges will be responsible, should we be surprised that cynicism about democratic institutions grows? . . .
>
> Of course, impassioned minorities and majorities will, from time to time, attempt to impose their passions on others, particularly those who are weak or unpopular. Protection of the right to be different, which is vital in a diverse society, is a proper judicial function. Ironically, this vital role of the courts is threatened by an overly assertive legal system.[71]

Commentators on contemporary politics present a gloomy picture. Senator Douglas indeed seems to have been a most prescient doomsayer. And, like his ancient predecessor Cassandra, it seems the Senator has gone unheeded. Even without the confounding factors of science and technology, the system is portrayed as inept at best. The ultimate loser, of course, is the electorate, whose membership is effectively disenfranchised long before the individual citizen even attempts to assimilate the impacts of the scientific revolution. Can it be that the American political system *accurately* conveys the impression of being run by special interests and a subverted press?

NOTES

Please note that all original footnotes and citations have been deleted from quoted passages throughout this book. To avoid unnecessary interruptions in reading, no ellipses replace these deletions.

1. Yang, *Chaotic Congress: Lawmakers' Inability to Clean Up Gridlock Is a Rising Frustration*, WALL ST. J., Mar. 21, 1988, at 1.
2. Wengert, *Environmental Policy and Political Decisions: The Reconciliation of Facts, Values, and Interests*, in POLITICS AND ECOLOGY (P. Foss, ed. 1972), at 45.
3. U.S. Constitution, Article I, Section 8.
4. The Bill of Rights, Article X.
5. U.S. Constitution, Articles I, II, and III.
6. Ibid., Article VI, Section 2: "This Constitution, and the Laws of the United States which shall be made in Pursuance thereof. . . , shall be the supreme Law of the Land; and the Judges in every State shall be bound thereby."
7. Ibid., Article I, Section 8: "The Congress shall have the Power . . . To regulate Commerce . . . among the several States."
8. *See* the Bill of Rights, Articles I, IV, V, VI, IX, X.

9. R. LITAN and W. NORDHAUS, REFORMING FEDERAL REGULATION (1983), at 32.

10. W. LUNCH, THE NATIONALIZATION OF AMERICAN POLITICS (1987). William M. Lunch was writing from the Political Sciences Department at Oregon State University.

11. Ibid.

12. Ibid., at 20–21. According to Lunch, television itself has fundamentally altered political communication. By moving presidential nomination and politics from the smoke-filled rooms to the television screen, the medium has provided more candidates an opening to nomination. TV ads are now essential to campaigns, because these ads establish a candidate's image. Moreover, the field narrows to two contenders rapidly because of the horse-race perception: Two are easier to follow than many. Indeed, "now more than ever, *perception is reality in politics*." Ibid., at 52 (emphasis added).

For more on American affluence as a factor in the evolution of environmental protection initiatives *see* W. TUCKER, PROCESS AND PRIVILEGE: AMERICA IN THE AGE OF ENVIRONMENTALISM (1982).

13. Lunch, *supra* note 10, at 5.

14. Ibid., at 6–7.

15. Ibid., at 125–128, 264.

16. Ibid., at 165–196 and *passim*.

17. Litan and Nordhaus, *supra* note 9 *passim*. Economic regulation controls entry into particular lines of business, sets prices, and establishes standards of service. As economic regulation becomes antiquated, deregulation follows. Social regulation seeks to correct the side effects and externalities of economic activity. The effect of economic regulation can be measured in dollars, but the ramifications of social regulation are never particularly amenable to measurement.

18. For a comparison of costs and benefits among programs of social regulation, *see* ibid., tables, at 11 and 15.

19. Ibid., at 18–19 and tables at 22 and 23.

20. Lunch, *supra* note 10, at 152.

21. Ibid., at 153, 163.

22. Litan and Nordhaus, *supra* note 9, at 43–44.

23. Lunch, *supra* note 10, at 172–183, *passim*.

24. Ibid., at 184.

25. Ibid., at 191.

26. Ibid., at 197.

27. Ibid., at 214.

28. Litan and Nordhaus, *supra* note 9, *passim*.

29. Ibid., at 103.

30. Lunch, *supra* note 10, at 134.

31. Ibid., at 151–153.

32. Ibid., at 131–164, *passim*. Since 1953 the U.S. Supreme Court, according to Lunch, has been engaged in an unprecedented era of reform and activism. There was moderation in the trend from 1969 through 1986, but no reversal.

33. Ibid., at 133. Lunch uses the word "delegate." A more critical commentator might find "relegate" more accurate.

34. Ibid., at 138. In this context Lunch observes: "The Supreme Court has pe-

riodically fallen under the spell of the dominant intellectual bias of its time and then made an effort to read that view into the Constitution. If there is any place . . . where ideology has . . . been almost directly translated into public policy, it has been the Supreme Court." Ibid., at 138.

35. Ibid., at 154–155.

36. Ibid., at 158. Others may be less candid about any such personalized interest, purporting to represent the public at large, without benefit of any known mandate.

37. Ibid. An additional question, and one raised by Lunch only by implication, is, who has determined that the interest represented by the self-appointed advocates itself represents any consensus of the public at large. *See also*, e.g., Frederickson, *The Recombinant DNA Controversy: The NIH Viewpoint*, in THE GENE-SPLICING WARS (R. Zilinskas and B. Zimmerman, eds., 1986), 13; and chapters 7 and 8, *infra*.

38. *See* Lunch, ibid., at 148.

39. Frederickson, *supra* note 37.

40. P. DOUGLAS, ETHICS IN GOVERNMENT (1952, 1972 reprint), at 32.

41. The discussion to follow in the text is from the prophetic work of Senator Douglas, ibid. Although published in the early fifties, this work was reprinted in 1972 and might easily have been written today.

42. Ibid., at 29.

43. Ibid., at 30.

44. Ibid.

45. Litan and Nordhaus, *supra* note 9, at 48.

46. Douglas, *supra* note 40, at 11, 27.

47. Litan and Nordhaus, *supra* note 9, at 49–103, *passim*.

48. "The expansion of national government has . . . created new opportunities for constituent services. From grant procedures for the National Endowment for the Arts to security classifications for small business, there are now federal rules and regulations baffling to most citizens but increasingly inescapable." Lunch, *supra* note 10, at 115.

49. Ibid., page 130.

50. Douglas, *supra* note 40, *passim*.

51. Lunch, *supra* note 10, at 163.

52. Ibid., at 151. Lunch asserts with some justification that one basis of extensive ambiguity in recent law is found in the difficulty in assembling needed coalitions simply to pass a bill. He expresses a reality difficult to accept, but all too accurate: "If everyone voting for a given bill was required first to understand exactly what it meant, nothing would ever pass." Ibid., at 151.

If members of Congress have abdicated the responsibility to comprehend the detail of the very laws they are passing, one is faced with the question: How is the electorate to comprehend the issues? And without comprehension does the participation of the electorate in policy formulation not become a sham, a de facto disenfranchisement?

53. Ibid., at 172.

54. Litan and Nordhaus, *supra* note 9, at 33.

55. Douglas, *supra* note 40, at 30 (emphasis added).

56. Lunch, *supra* note 10, at 147.

57. Koshland, *Judicial Impact Statements*, 239 SCI. at 1225, Mar. 11, 1988 editorial. Daniel E. Koshland, Jr., is the editor of *Science*.

58. Lunch, *supra* note 10, at 117.

59. Ibid., at 133–138 and 197–223 *passim*. "Special interest" is a broader and probably a more accurate designation than "public interest."
60. Ibid., at 223.
61. Ibid., at 193.
62. Yang, *supra* note 1.
63. Nordhaus and Litan, *supra* note 9, at 2. The authors expand on the observation:

> Missing the forest by watching the trees caused little waste when the regulatory agencies were few and their missions modest. In recent years, however, the scope and impact of government regulation have proliferated. Depending on how they are counted, there were fewer than ten agencies charged with regulatory responsibilities prior to 1900; today the number stands at over eighty. This growth in the number of regulatory agencies is reflected in the substantial private-sector impact of federal regulation. . . . [G]overnment regulations in the aggregate now require private firms and individuals to spend upward of $50 billion *each year*. The benefits they provide, although more difficult to quantify, appear to be of a similar magnitude.

64. Ibid., at 3.
65. Ibid., at 3–4.
66. Yang, *supra* note 1.
67. Ibid.
68. Douglas, *supra* note 40, *passim*.
69. Lunch, *supra* note 10, *passim*.
70. Ibid., at 163.
71. Ibid.

CHAPTER 2

Policy Formulation at the Intersection of Law, Science, and Technology

THE NATURE OF SCIENCE

> "The special dependence of science on the natural environment means that the content of scientific knowledge is determined, in principle, by forces or conditions which are beyond human control."[1]

Science can be defined as the systematic knowledge of the physical or material world or as a branch of knowledge or study dealing with a body of facts or truths systematically arranged and showing the operation of general laws.[2] Technology in turn is that branch of knowledge which deals with applied science, industrial arts, engineering, and the like. Finally, engineering is the art or science of making practical applications of the knowledge of pure (basic) sciences, such as physics, chemistry, and biology.[3]

These bland dictionary definitions are accurate so far as they go, but they leave much unsaid. More important is what science and technology *do* and how they go about it.

Essential to science as a unique pursuit of knowledge is the *scientific method*: principles and procedures for the systematic pursuit of knowledge involving the recognition and formulation of a problem, the collection of data through observation and experimentation, and the formulation and testing of hypotheses.[4] Each methodical step is, as the legal profession would put it, *necessary* to science, but none in and of itself is *sufficient*, at least in the long run, to constitute science. While any scientist at a given point in time may well engage in something less than the whole method, science depends on all three steps to be science. Technology, then, is a scientific means for achieving a practical purpose. As such it reflects the totality of the methods employed to provide objects necessary for human sustenance and comfort.[5]

A more provocative definition of science separates that human endeavor from those identified as philosophy, religion, technology, or magic. Modern science is the intellectual endeavor that depends entirely on the assumption that nature is operating according to general laws hidden but discernible through systematically controlled observation and experimentation.[6]

Although definition is a prerequisite to addressing the nature of science, it is the claims of science which are of more immediate concern to society. Stated with characteristic bluntness, science is universal in its claims.

Conceptually, science embraces everything from inanimate nature through life to human nature and human society. Functionally, science includes the interrelated processes through which we continuously acquire knowledge of the natural world. Scientific method proceeds from prior knowledge in constructing general theories or systems from which testable hypotheses can be derived. Testing occurs through quantifiable observations under controlled conditions. Through its acquisition of knowledge of the natural world, science moves first toward understanding nature, then towards prediction of natural events and, ultimately, into control of natural forces.[7]

Many a philosophical debate centers on the nature of science and its evolving relationship to philosophy, religion, or magic, but such debates are beyond the scope of this effort. That the place of science in our lives is in revolution and therefore subject to major debate is, however, the second of two primary contexts for the present discussion.

Significantly, contemporary science is self-reinforcing. That is, lines of enquiry arise primarily within the context of scientific method itself. Hence, science no longer depends on its societal environment for stimulation and guidance, despite its financial dependence on society. Answers to scientific enquiry merely raise new questions in series without any termination point that is presently perceivable. As a result all scientific concepts and even laws bear an intrinsically tentative status. As science evolves, old systems are displaced or are replaced outright by new.[8]

This aspect of science and technology may be the one that most disturbs the lay observer. When some feature of science or technology touches our individual day-to-day life, uncertainty and the prospect of unending change are simply unacceptable to most of us.

This is also the point at which the individual directions taken by scientists must be distinguished from the progress of science at large. Though the former may be subjected to some control, the latter is virtually uncontrollable because the course of scientific development is inherently unpredictable. No one can tell how nature is going to answer future scientific questions or even what the questions themselves will be. In essence, science can be likened more to an adventure than to an institution. This, moreover, is an adventure in which the whole of society takes part. Scientific progress is neither a stable institution nor a stabilizing influence on other institutions—quite the contrary. Indeed, a society emphasizing science will itself be dynamic, with drastic effects on all institutions.[9]

Science, then, is not merely uncertain and unpredictable. It is beyond the control of the society from which it is emerging and which it is presumably to serve.

If this is indeed the nature of science, the sociology of science becomes crucial to its accommodation by (and *to*) its host, society, that is, for purposes of the present discussion, the peculiar culture that is contemporary America. Sociological analysis could pave the way to directing science and, especially, technology, in lieu of actually exerting control over one or the other.

In order to come to grips with science, it is first necessary to gain some comprehension of how scientists interact with their work and with their colleagues. Then it is appropriate to consider the corresponding relationships with society at large.

The Society of Scientists

> "We need to examine the internal concepts of science to find out how we might analyze the connections of science with values. We need to study more thoroughly the importance of genetic evolution, physiological structure, and psychological conditioning for understanding our social behavior and values."[10]

Scientists engage in a major personal commitment in their professional endeavors. Science is something they *want* to do; a source of enjoyment. The layman may find it difficult to accept scientific work as creative, yet each contribution must be unique, a characteristic shared with works of creative artists. Nevertheless, science differs from most creative activities in its professional organization. Creativity in science is uniquely due to its organization, and, uniquely, it is from within that scientists are evaluated by their peers.[11]

As has been suggested, science at large is a process that transcends individual scientists, research teams, and even institutions. The actual outcome of research efforts may not have been the specific goal originally set. Some directions and some goals emerge more by accident than by design.[12]

Another unique feature of science is the absence of formal licensing or educational requirements. Entrance into the scientific community is determined exclusively by the community itself. Thus, scientists have no clients as such and are bound by no formal code of ethics in their relationships to nonscientific sponsors. The closest analogy to a client is the scientific community itself, which has no need of the protection against malpractice that is required for the clients of other professionals.[13]

In short, there is a profound difference between science and institutions more familiar to the public at large. Each society's characteristic religious, educational, political, and economic institutions will be more or less developed. These institutions are generally recognized and incorporate latent functions. Relatively changeless, they may exhibit gradual alteration in performing their respective functions without disrupting societal stability. Science, however, has emerged on a large scale only in Western civilization and in recent centuries, and it is an institution more disruptive than integrative. Through science, Western society has been drastically transformed. Enormous benefits and disastrous results are perceived in the transformation, a situation unlikely to foster a stable relationship between science and society.[14]

In contemporary America scientists are seen, even by other scientists, as an isolated and almost a cloddish faction. By way of particularly pertinent example, "pure scientists," past and present, have been justly accused of a devastating ignorance of productive industry. Scientists and their technological kin, engineers, often are both in total misunderstanding; but only the engineers have managed to present a disciplined face to the world. Scientists, it might be asserted, can be dimwitted even about applied science.[15]

Ultimately, for a scientist success is a source of esteem. But a scientist's success may have a definition the layman will find difficult to accept, especially when he or she has a personal involvement with the situation at hand:

> Competent evaluation of professional work requires that the effects of this work be distinguished from situational effects beyond the professional practitioner's control. The importance of such a distinction is illustrated by the possibility that a physician who saves the life of only one patient out of one hundred may thereby be performing excellently *if* he is treating a disease which is ordinarily survived by only one out of one thousand of its victims. *Only professional colleagues can be counted on to make proper allowances for situational factors, and hence only they can be considered generally competent to evaluate professional activity.*[16]

Hence, it would appear that nothing within the scientific culture bodes well for a functioning relationship between the sciences and contemporary American society.

It is the final sentence quoted above, however, that may represent the major impediment to effective interaction at every point where public policy intersects with science or technology. Here, in one succinct sentence, is the crux of many of the problems arising when law, science, and society meet. Legislators, regulators, members of the judiciary, jurors, and even other scientists addressing the same issue may be laymen and, hence, the bearers of questionable credentials for addressing the matter at issue before them.

Science and Society

> "The rate of technological advancement is far faster than our political, social, and economic institutions can adjust to in such a rapidly changing world."[17]

Basically the sociological view of science finds it to be more a cultural process than an institution of society. There are three sociological perspectives on science: as the cultural counterpart of individual cognitive development, as an outgrowth of traditional cultural knowledge, and as a cognitive form of cultural development.[18]

Whenever science is introduced, it is wise to keep in mind that there is a hazard, born of ignorance. Individuals commonly fill in gaps in our observational (empirical) knowledge. Too often the gap is filled through imaginative speculation, the interpretations of which become a part of culture. We must, however, be prepared to accept that, while these speculations may substitute for correction, they are not true corrections. Speculation firmly entrenched in prevailing culture is instead a major obstacle to legitimate processes of correction.[19]

It is suggested, then, that both sociologists and individual members of society have a distorted view of science, simply because those observing are outsiders. Presumably, the further an observer is removed from the subject the greater the distortion. And yet, as will become increasingly clear, the electorate, however distorted its view may be, would make demands of science and, when the response is somehow unsatisfactory, alter science to suit those demands. The representatives of the electorate, in their turn, can be expected to incorporate unavoidable speculation in the course of formulating policy. Speculation, then, is the second major impediment to effective interaction between society and science. Speculation will be almost impossible to excise, and scientists who make the attempt are not likely to be well received.

Unfortunately for all concerned, understood or misunderstood, underlying scientific reality can not itself be altered, no matter the need, no matter the lofty goals. Ultimately, while we can encourage or inhibit science, we as a society can neither predict its course nor unilaterally shape it over the long range. Yet corresponding technological implications of those courses may represent our own transformation.[20]

For all of societal encouragement or discouragement of science, it is fundamentally impossible for us to determine, to tailor to our interests, scientific knowledge itself. Natural phenomena, the contents of scientific knowledge, are to be observed, not recreated in an image we prefer. To the extent we seek to intervene to require a certain scientific formulation or to prohibit one, in disregard for scientific revelation, what results is something quite distinguishable from science.[21]

Given the existence of so unwieldy an amalgamation of ignorance, isolation, and immovability in the face of societal needs and fears, both scientists and society are in for some substantial battles and considerable frustration. Furthermore, as we shall be exploring, science is one of two disparate cultures in the Western world.

At one pole, the scientific culture really is a culture, not only in an intellectual but also in an anthropological sense. That is, its members need not, and of course often do not, always completely understand each other; biologists more often than not will have a pretty hazy idea of contemporary physics; but there are common attitudes, common standards and patterns of behaviour, common approaches and assumptions. This goes surprisingly wide and deep. It cuts across other mental patterns, such as those of religion or politics or class.[22]

The effect of this basic alienation is only exacerbated by the recent emergence of large-scale science, the timing and scale of which preclude not only perspective but also definitive studies of its long-range effect on society.[23]

It requires little imagination to foresee major problems as a scientific revolution arises from the more stately evolutionary pace, especially as science becomes entrenched. Science is exerting its own pressures on the character and organization of human endeavors. As science and production draw closer together, scientific and technological progress gathers speed, while natural resources may be used more and more rationally, from reclamation of land to the aspect of managing climate.[24] Science is, in fact, moving into every sphere of politics and of economic management.[25]

What must be recognized is that science (like the scientist) is represented by white and black and shades of gray. When it comes to applied science, it is one thing to escape associated dangers. It is even more difficult and demanding of society, but more enriching, to accomplish the good that applied science places within our power. That accomplishment demands energy and self-knowledge along with new skills. But using science in any application also requires perceptions into politics, closed as well as open.[26] How are these needs to be met?

Changes in education will not be enough, as C. P. Snow predicted. But, as he warned, without changes in education we will not even recognize what the problems are. That cultural division, of his time and into our own, makes us even more obtuse than we need to be. It is essential that our best minds be alerted.[27] It is equally essential that the electorate be aware!

Neither will be an easy task, for at that other pole, distanced from the science culture, where there is admittedly a wider spread of attitudes, there is also total incomprehension of science. From incomprehension flows the unscientific, which in turn gives rise to the antiscientific. Scientists are said to have the future in their bones, but cultural traditionalists are more likely to respond with the wish that the future did not exist. Our own culture is managing our world in this ignorance of science.[28]

Under these circumstances and given what we have learned of contemporary American politics, it is more daunting than reassuring to find science compared to bureaucracy. Nominally, neither institution is subject to the caprice of personality. Similarities in the two should facilitate assimilation of scientific research into bureaucratic settings. Yet, in fact, bureaucratic settings potentially threaten scientific freedom in the selection of research topics, interpretation of results, and communication of ideas and findings among the scientific community.[29]

Maurice Richter was concerned about the effects of politics on science. What we face today is the inability of the two systems, too often persistently parallel when convergence is in order, to communicate effectively with each other. Unless eliminated or neutralized, the two impediments, the inability of any nonscientist to judge science accurately and speculation as a poor substitute for correction, will always confound communication. Poor communication can only result in policies derived from scientific and technological gibberish.

THE REVOLUTION AND THE TWO CULTURES

> "The cautious analytical approach is a discouraging tactic against uninformed fear, mysticism and political opportunism."[30]

The Scientific Revolution

> "The powerful new methods are here. Applying them may well require the use of brain enhancers."[31]

The existence of a revolution in science and technology cannot be doubted. A single glance at the daily headlines or through the weekly news magazines will substantiate that we are presently experiencing scientific advances at a rate and of a scope without historical precedent. But what is being reported tends to be science in the abstract. What renders the abstract real to the individual participating in the contemporary American political scene is better explored through specific circumstances.

Among the most notorious of these are the politics of genetics in the thirties and the ongoing debate surrounding Agent Orange. Both will be considered briefly as a preface to expanded consideration of the relationships of science and technology at their merger with policy formulation; an examination of that merger in national environmental policy will follow. From those starting points, issues of pharmaceutical innovation and the matter of biotechnology, a technological revolution in and of itself, will provide detailed illustrations of the blessings and of the problems, possibly intractable, that attend policy formulation at its intersection with innovative science or technology. The place of the individual in these processes will become increasingly apparent.

In the sixties we had only begun to live with the industrial *cum* scientific revolution and had taken steps to control the industrial element and to see beyond its advances to acknowledge its external costs. The scientific element remained beyond our imaginative understanding. To our recent ancestors the industrial aspect of this dual revolution was an improvement over the past but needed to be made better.[32] They, and we, may not be so amenable to the scientific aspect of the revolution.

Sometimes only an outsider can lend the proper perspective to "local" events.[33] A Russian commentator of the seventies noted the unparalleled developments in science and technology originating at the turn of the century. He observed that an amazing 90 percent of all the scientists in history are our contemporaries. Those unprecedented advances, of which manned flight is but one example, are occurring within periods shorter than a single lifetime. To him, however, the advances constituted a social and economic factor of the utmost importance.

From the socialist vantage point, the scientific revolution is more than background; it is the means by which the world is made a better place for people, an instrument of social progress. Significantly, applied research, itself novel, further accelerates progress arising from influences of science and industry upon each other. Closer ties between modern science and technological progress leads to rapid transition from knowledge of "deeper" laws of nature and the production of useful ends.[34]

Progress in science and technology itself leads to major changes in professions and skills, calling for a knowledgeable, well-trained work force. These changes can result in the ouster of the existing work force as redundant. Both older workers and young people can be displaced, the former at a dead end, the latter unable to gain proper training for entry. Future prospects will be grim, and additional repercussions can follow.[35]

The demand of the revolution in science and technology calls for better-educated employees, the education they need may be beyond the means of many, especially any who must leave school to earn a living. Those who cannot gain needed education will be doomed to unemployment[36] or underemployment.

It is not the intent of this effort to cast aspersions on any political philosophy as such—or to counter any such aspersions. But in considering our circumstances we would be well advised to ponder the foregoing commentary (the rhetoric of which has been tempered as inappropriate in the present context and in light of recent events) and to determine for ourselves the extent to which a word of valid warning has been imparted. It is certainly worth noting that, to these Russian commentators, advancing technology is a boon to serve the political philosophy. It is neither a bane to be regulated out of existence nor something subject to alteration according to political whim.

It is difficult to argue against these two propositions: (1) The scientific and technological revolution has led to a profound awareness of the enhanced role of science in society. (2) The elevation of guidance in science and science-training to a state policy has been developing since World War II.[37] Certain corollaries to these propositions will be the ultimate focus of this work.

Science Versus Humanities

> "How is it possible to consider someone as educated who knows virtually none of the wealth of new scientific principles accumulated since, and before, 1937?"[38]

C. P. Snow has reflected on the nature of the antipathy perceived between the world of science and that of the humanities. Perhaps the most salient feature of the alienation is the mutual degree of ignorance, an ignorance no society can afford to foster. That ignorance is widespread even among those believed to be highly educated. While the lapse may be mutual, in the present context ignorance of science, illustrated by the arrogant inability to describe the Second Law of Thermodynamics, is somewhat more damaging.[39]

Ignorance of science is hardly unique to contemporary American culture. Even though the agricultural and industrial–scientific revolutions "are the only qualitative changes in social living that men have ever known;"[40] it has been alleged that traditional culture never noticed them, or having taken notice, did not like what it saw. As a result, it can also be alleged with some justification that the majority of the most clever in the Western world have as much insight into basic physics, for example, as did their neolithic ancestors.[41] Amusing to be sure, but a sad commentary. The result:

> The clashing point of two subjects, two disciplines, two cultures—of two galaxies, so far as that goes—ought to produce creative chances. . . . The chances are there now. But they are there, as it were, in a vacuum, because those in the two cultures can't talk to each other.[42]

Scientists have no reason to be smug either. Among too many of them applied science is something to be relegated to the second-rate mind. Too many are inflicted with a pride in doing scientific work that "could not, in any conceivable circumstances, have any practical use."[43]

Nonscientists remain the overwhelming majority, even in societies, such as ours, that place great emphasis on science. This vast majority has neither much understanding of science nor much interest in participating in the revolution.[44] On a personal level it may be harmless enough to reject the revolution. But there is harm in imposing, even through passivity, such a choice on others who are not free to choose the concomitant lack of food, infant deaths, and the like.[45]

With or without participation by the vast majority of the electorate, large-scale science can be expected to continue into the future. As a result, one of three eventualities can be expected: Science could continue in its uneasy, unstable relationship with society along with a continued disruptive impact on major social institutions. On the other hand science could become clearly institutionalized as a societal subsystem. Then the relationship would be stabilized. Science would overtly serve its specialized function for society and enjoy societal support in return. Obviously, there has been some impulsion to that end. The third eventuality is the elimination of the isolated, specialized character of science as a small segment of society. Science would then become a major concern of society as a whole, with science for its own sake a major societal goal.[46] We are also headed, without benefit of much steering, in that direction. It is this direction which has the best chance of eliminating the impediments of parochialism and speculation.

Whatever eventuates, it must be abundantly clear that a lack of communication between two cultures, whether unavoidable or a matter of attitude, is dangerous when science has become determinative of our destiny. Science is literally a matter of life or death. It is imperative, then, that those who formulate policy know bad scientific advice when it is proferred. Knowledge of available potentials limited to scientists is intolerable. We need it to fulfill social aspirations and to avoid disasters.[47]

EUGENICS: YESTERDAY'S CASE IN POINT AND PRELUDE

> "[I]n given historical situations, such as those in . . . Germany and . . . Russia in the 1920's, rival scientific theories about human nature have invariably arisen within the context of given social and political circumstances, of competing political ideologies and economic motives, and of existing systems of technological capabilities."[48]

The external factors imposed on science by cultures instill differential value implications in rival scientific theories. Thus, whenever we ask whether a matter of science contains inherent social or political values, we are asking the wrong question. Instead we should be examining links between science and the values that are contingent upon the existing system of politics, social situations, and technological capabilities, in conjunction with the persuasiveness of ideologies.[49]

Because every scientific theory and every technological innovation always arises in such a milieu, value-free interpretation of the theory or innovation can follow only upon severance from that milieu. Indeed, history continually demonstrates just how massive the momentary influence of these external forces on science and technology can be. Germany and Russia are cases in point.

The science was genetics, more precisely, the technology that was becoming associated with eugenics. Germany and Russia reflected opposite ends of the "nature versus nurture" controversy of the time. The applications were human eugenics and, in Russia, agricultural genetics. Significantly, the known biological science and the tested hypotheses were the same in both countries. Yet, neither Germany nor Russia closed the logical circle. Instead, each focused on an extreme interpretation.

The horrific results of the actual practice of human eugenics in Germany are notorious and need not be recounted here. It is sufficient to understand that they went far in discrediting eugenics throughout the world. The theory may have been correct and the applications scientifically sound, but the associated values were a gross perversion. On the other hand, in its zeal to embrace theory philosophically acceptable in human application, Russia expressly renounced eugenics as "bourgeois doctrine." But application to agricultural practices of a scientific theory in political favor actually did the Russian people far more harm than good.

The situation in the two countries was unique in that each was emerging from monarchy. The political situations, then, may well have precipitated the respective correlations between scientific theory and political worldview. The science itself was too immature to provide technological resolutions to the value controversies, had they been sought. Between them, Germany and Russia managed to prove that neither sound science nor human values can be allowed to take complete precedence over the other without the risk of devastating effect.

While the German experience was notorious worldwide, that of Russia was more insidious but still very damaging—to science and technology, for a while, but ultimately in humanitarian terms as well. For this reason, the Russian experience is the more effective lesson for application in contemporary American society.

> Lysenko's theory . . . ,based on the inheritance of acquired characteristics, assigned an overwhelming weight to the nurture side of the equation. His naively optimistic theory of inheritance granted a much greater degree of control than the results of researches in biology had indicated was possible.[50]

In Russia, eugenics came to an end with the emergence of Lysenko's theory. Theory was subjected to Soviet political loyalties, historical associations, and dedication to internal agricultural policies. Neither scientific validity nor logical rigor played much part in determining the results. In demanding the approach best suiting its political theory as applied to the human race, Russian policy did substantial damage to the agronomy needed to feed its people.

The argument, resorted to by those scientists who attempted to support classical genetics as socialistically superior to Lamarckian heredity, was not sufficiently persuasive to stem popular belief in the inheritance of acquired features. In short, Lamarck's theory was more congenial to the idea of creating a new society and a new culture than was Mendelian genetics. For one thing, as the contrived argument was fairly sophisticated, it was not heard or, if heard, was ignored. Veracity notwithstanding, *simple arguments won out over subtle ones.*

> Societies have thrust upon them scientific theories and technological innovations which may either buttress or counter existing value preferences. The powerful groups in society are usually more successful than the weak ones in turning science and technology to their advantage, so there is a natural tendency for science and technology to buttress the values of the ruling classes and political groups.[51]

Clearly, the ultimate value of science does not lie in what it does for scientists. Its value lies instead in what it does for the intellectual and material assets (and liabilities) of society. The ways in which society is affected by science and society are ambiguous and can be disturbing. There is legitimate public concern with regard to leaving these endeavors entirely in the hands of the scientists and engineers. Nevertheless, history has demonstrated that it can also be disastrous to afford lay groups control over determinations of the inherent value of fundamental research.[52]

The major lesson to be learned through the Russian experience is that political favor may impose its priorities on developing science, at least for a time. In the long run, however, the scientific reality will prevail. Every culture must accept that fact. In doing so, it is essential to remember the grave toll that momentary, and quite illusory, political power (whether motivated by values benevolent or malevolent) can inflict before reality can break through. Every effort, then, should be directed at the elimination of at least the fallacies associated with the momentary political supremacy.

AGENT ORANGE: TODAY'S CASE IN POINT

> "I expect that the results from scientific studies will become of little more than academic interest unless they find overwhelming evidence of human damage. By the time results are available, many of the important decisions will have been made. A number of them have already been made."[53]

Probably the most difficult scientific reality for the layman to accept is the absence of scientific or technological answers to their most immediate and intensely personal questions. The course of science, contrary to popular belief, is effected neither by political demands nor by a heavy influx of monies. Neither means can force out answers scientific methods cannot otherwise provide on a timely basis. This is the situation with Agent Orange, which has become notorious without benefit of some underlying innovation in science placing human values at risk.

The immediate context is the Vietnam War combined with the current state of toxicological knowledge, in general and specifically with regard to the defoliant Agent Orange and its contaminant, dioxin.[54] The problem presented by the context is that science simply does not know the specific causes of most cancers, heart attacks, and other conditions. Moreover, science cannot tell us whether Vietnam veterans would be free of those conditions had they not been exposed in the course of the war to Agent Orange. Yet for us to assert on that basis that none of those conditions suffered by Vietnam veterans is ascribable to Agent Orange may be going too far. We simply do not know the answers.

Where once humanity turned to evil spirits and witchcraft in search of unavailable answers to disease, we now expect physical causes. When confronted with the unexpected or when there is little information as to cause available, we point to the environment, with special reference to anthropogenic chemicals.[55] The satisfaction in either case is in providing some reason for the affliction while absolving from responsibility the person harmed and conferring responsibility for control on some external agency.[56]

In the late seventies the Department of Defense rejected out of hand any association between Agent Orange and the afflictions being reported. This decision prompted the General Accounting Office (GAO) of Congress to initiate a study of the case. The resultant report concluded that one essential step was possible: *some* likely exposures to Agent Orange *could* be determined. The corresponding recommendation was to establish whether a study of health effects was needed.

In response, Congress mandated such a study by law and granted the Veterans' Administration six months just to plan it. At the same time the White House formed a working group to coordinate Agent Orange policy. It soon became clear that dioxin must be studied along with Agent Orange, and lengthy delays in the study plan arose out of a series of additional confounding factors. Eventually these confounding factors were held to be impossible to sort out, in effect rendering scientific analysis meaningless.

Instead of studying health effects, a science panel recommended a study of the Vietnam experience as a whole. The law was amended accordingly and, after considerable discussion, three groups (cohorts) were designated for study: Vietnam veterans apparently exposed to the chemical at issue, Vietnam veterans presumably not exposed, and veterans who had not served in Vietnam. These and related preliminary discussions continued into the eighties!

A full-fledged plan of study was not established and approved (with a budget of $70 million) by the powers-that-be until early 1984. Study materials and methods were scheduled for release in 1987, but actual results would not be available before 1989 and 1990. Even so, many uncertainties remained unresolved, throwing considerable doubt on the validity of whatever results might finally be obtained.[57]

In the meantime the matter went to court and was resolved not by litigation but by settlement. The scientific realities remain elusive despite pressure for resolution from all sides. As more data have been analyzed, the promise for resolution recedes. The study itself should be reevaluated according to the probability that meaningful answers will be forthcoming. If the probability is low, the study is pointless. Michael Gough saw little evidence of any such reappraisal. Moreover, even assuming a finding of no health effect,

> people who think the result is wrong can say that the exposure classifications are so likely to be off that the study was flawed. And they are right. Since there is no way to do a better job in classifying exposure, it might be wiser to consider stopping the study. And it might be more responsible to say that exposures were so low that no effects would be expected.[58]

Nevertheless, it is a foregone conclusion that, under public pressure, the study will proceed, while the issue as an item of public policy will fade away. It will remain real only for the veterans and their families who believe themselves to have been harmed.

Michael Gough poses a question of immediate interest. How can public perception depart so far from scientific realities at every level: theoretical, demonstrable, practical? What are the chances of bringing perception closer to reality? His answer is that the two will converge as the controversy recedes, as is suggested by more recent reporting and editorial stances on Agent Orange. They are more balanced than was the case in the initial stages of the controversy.

With respect to court action, Gough observes, "courts are generally considered a poor place to decide scientific questions; only the rare judge or juror has technical training to make an informed judgment."[59] To the extent legal action is deemed unsuccessful, media coverage will recede. As it does, balance and thence science and ultimately the public at large will all be served, because there will be fewer opportunities for the airing of extremes through television presentations and debates among vested interests. Though all could be educational, seldom is any guidance provided for the observer as to which view is generally accepted as more credible among scientists.[60] Significantly,

> [a]s opportunities to expound on dioxin's toxicity diminish because of fewer publicized court cases and congressional actions, it may appear that some common ground

between the two extremes is being reached. [Gough] doubt[s] that agreements will be reached; instead, the arguments will disappear from public view.[61]

Before the arguments can disappear, one final observation by Gough must be included for future reference throughout this volume. With his belief that the results of the ongoing study will prove to be negative, he offers rather a cynical prediction:

It is probable that as each [segment] is released, the principal investigator will be interviewed, state his or her conclusions, and be shunted off-camera and replaced by a proponent of the view that dioxin has caused many adverse health effects. . . . [T]he news media, in making . . . balanced presentations, project anecdotal accounts as somehow equal in structure and validity to large, well-conducted studies. In leaving the viewer to choose between the negative study and the proponent's views, it obscures the emerging consensus about the absence of demonstrable effects of dioxin on human health and, I think, downgrades the contributions of scientists who conduct honest, credible studies.[62]

Contemporary American society is, in effect, proceeding on two tracks, in regard to this and any number of other issues based in innovative science or technology—or simply based on scientific unknowns. Unfortunately the two tracks will be afforded slight recognition as they proceed independently. Science will say the jury is still out, while judges and juries (and other formulators of policy) decide the issue anyway. While the scientists are still searching out the relevant facts, the formulators of policy will evaluate the available scientific evidence, will make their decisions, and will assign responsibilities.

As the bulk of this volume will illustrate in other contexts, neither those who believe themselves injured by Agent Orange nor the public at large will in the end be served. Furthermore, the ongoing study and many others like it constitute, in effect, a scientific sham, perpetrated solely for political reasons at the possible instigation of purportedly disinterested parties. The sad irony is that any real underlying public demand leading to such political efforts is not being served either.

There is a saying among jurists that hard facts make bad law. There should be a comparable saying to the effect that political expediency cannot create science; good, bad or indifferent. Instead, what political expediency is more likely to accomplish is the perversion of science to the detriment of the host culture, a fact to which history has so forcefully attested through the recent experiences of Germany and Russia.

NOTES

1. M. RICHTER, SCIENCE AS A CULTURAL PROCESS (1972), at 6. Maurice N. Richter, Jr., intended his work in part for the general public concerned with

the relation of science to society. Ibid., at ix. This concern is the major focus of this book. Something of the import of this effort is suggested by the contemporaneous observation of the vice president of the USSR Academy of Sciences to the effect that scientific and technological development is a major sector of competition between capitalism and socialism. Millionshchikov, *The Crucial Test for Mankind*, in THE SCIENTIFIC AND TECHNOLOGICAL REVOLUTION (R. Daglish, ed., 1972), 13. This anthology of essays was published in Moscow. There was no identification of any translator into English.

2. THE RANDOM HOUSE DICTIONARY OF THE ENGLISH LANGUAGE (Stein, ed. in chief, 1971 unabridged edition).

3. Ibid.

4. WEBSTER'S NINTH NEW COLLEGIATE DICTIONARY (Mish, ed. in chief, 1983).

5. Ibid.

6. Richter, *supra* note 1, at ix.

7. Ibid., at 14.

8. Ibid., at 64–83, *passim*.

9. Ibid., at 18.

10. L. GRAHAM, BETWEEN SCIENCE AND VALUES (1981), at 382.

11. Richter, *supra* note 1, at 28–31, *passim*.

12. Ibid., at 15–16.

13. Ibid., at 29.

14. Ibid., at 18–19.

15. C. SNOW, THE TWO CULTURES: AND A SECOND LOOK (Cambridge University Press ed. 1969), at 31–32. The celebrated C. P. Snow was a scientist by training and a writer by vocation. He spent his working hours with scientists and his evenings with writers. From these situations arose his perception of the two cultures. Even today, those who would become initiated into a professional circle must be indoctrinated in its secret language—what has come to be known as jargon.

16. Richter, *supra* note 1, at 27 (emphasis in the last sentence added).

17. B. ZIMMERMAN, BIOFUTURE: CONFRONTING THE GENETIC ERA (1984), at 6.

18. Richter, *supra* note 1, at 33.

19. Ibid., at 38.

20. Ibid., at 127–128.

21. Ibid., at 6.

22. Snow, *supra* note 15, at 9.

23. Richter, *supra* note 1, at 6.

24. Trapeznikov, *Leninism and the Scientific and Technological Revolution*, in SCIENTIFIC REVOLUTION, *supra* note 1, 55, at 58. Sergei Trapeznikov holds a doctorate in history. This provocative anthology provides an enlightening perspective on law, science, and technology while offering some chilling food for thought in the context of the present work. *See*, e.g., Millionshchikov, *supra* note 1, at 27.

25. Trapeznikov, ibid., at 59.

26. Snow, *supra* note 15, at 99.

27. Ibid., at 100.

28. Ibid., at 10–11.

29. Richter, *supra* note 1, at 87.

30. Singer, *The Recombinant DNA Debate*, 196 SCI. 127, Apr. 8, 1977, editorial. Maxine Singer is a prominent figure in this controversy. *See* chapters 7 and 8, *infra*.

31. 236 SCI. 1157, June 6, 1987, editorial. The accelerating pace at which practical applications follow from revelation of scientific principles is brought home by Trapeznikov, *supra* note 24, at 57–58, with reference to such "ordinary" technologies as photography, telephones, television, and transistors.

32. Snow, *supra* note 16. *See also* Richter, *supra* note 1, at 53.

33. The following discussion is taken from comments of Millionshchikov, *supra* note 1, *passim*.

34. Ibid., at 14.

35. Trapeznikov, *supra* note 24, at 63. *See also* E. USEEM, LOW TECH EDUCATION IN A HIGH TECH WORLD (1986).

36. Dalin, *The Scientific and Technological Revolution and Aggravation of the Contradictions of Capitalism*, in Daglish, *supra* note 1, 173, at 176. Sergei Dalin holds a doctorate in economics.

37. Marakhov and Meleshchenko, *Specific Features and Social Consequences of the Scientific and Technological Revolution*, in Daglish, *supra* note 1, at 144.

38. Westheimer, *Are Our Universities Rotten at the "Core"?*, 236 SCI. 1165, Jun. 5, 1987, Policy Forum.

39. Snow, *supra* note 16, *passim*. The Second Law of Thermodynamics holds that all energy transformations in the universe proceed to greater dispersal of energy. This law governs all chemical reactions, including those associated with living systems. As C. P. Snow observes, recognition of this law is the scientific equivalent of familiarity with a work of Shakespeare.

40. Ibid., at 23.

41. Ibid., at 15. *See also* Richter, *supra* note 1, at 53.

42. Snow, *supra* note 16, at 16.

43. Ibid., at 32. C. P. Snow included himself among these isolationist-scientists. The attitude is still to be found in the halls of academia, although it may be more subtly expressed these days. It will be considered in greater detail in Part III of this book.

Furthermore, scientists are coming to accept their own failure to make their community aware of their accountability to the public. Alert scientists accept the impropriety of hiding behind academic regalia and isolating themselves from the people they serve. *See* Cairns, John, Jr., *A Survival Program for Taxonomists and Their Institutions*, 5 A. SYSTEMATICS COLLECTIONS NEWSLETTER 32 (June 1977).

44. Richter, *supra* note 1, at 20.

45. Snow, *supra* note 16.

46. Richter, *supra* note 1, at 97.

47. Snow, *supra* note 16, at 98–99.

48. Graham, *supra* note 10, at 253.

49. *Id.* The entire discussion in the text is taken from this work. While the science in question had emerged in the twenties, the ensuing discussions were *political*. The politics were of the thirties. The controversy itself reflects scientific extremes: Those who asserted the nature side claimed all characteristics depended on the genetics of the species at issue. Those who asserted the nurture side claimed total dependence on the environment in which the individual within the species exists. In fact, geneticists are learning that environment does influence genetically mediated charac-

teristics, probably within molecular limits unique to each organism, on both levels, that of the species and that of the individual.

50. Ibid., at 246.

51. Ibid., at 254.

52. Ibid., at 318.

53. M. GOUGH, DIOXIN, AGENT ORANGE: THE FACTS (1986), at 232. Michael Gough was directly involved in the events as a member of the Office of Technology Assessment (OTA) and as OTA's representative to the White House agency involved in the events.

This concern on the part of the commentator has been borne out in the report (referenced in the text) on the Vietnam experience. Public and official attention have been held by the Agent Orange issue, which may well be without sound basis, at the expense of issues of psychological damage, which may prove to be both substantial and unique to this war experience. Roberts, *Vietnam's Psychological Toll*, 241 SCI. 159, July 8, 1988.

54. Gough, ibid., *passim*. The entire discussion in the text is taken from this work.

55. Ibid., at 227.

56. For more on this propensity, *see* the discussion of universal, multilevel reaction to trauma in chapter 5. For more on unfolding events, *see* Hays, *Proposals from Federal Agencies to Ease Dioxin Standards Renew Debate on Risk*, WALL ST. J., Jan. 27, 1988, at 27.

57. Gough, *supra* note 53, at 95–100.

58. Ibid., at 102.

59. Ibid., at 222.

60. Ibid., at 223–224.

61. Ibid., at 226.

62. Ibid.

CHAPTER 3

Implications for the Real World

WHERE WORLDS COLLIDE: SCIENCE AND CONTEMPORARY BUREAUCRACY

> "Science is inevitably tilted toward future gain. The adjudication of priorities in the present... pales in significance compared to projects that affect... future generations."[1]

The foregoing explorations hold out little promise for any integration in the real world of the three cultures: scientific-technological, politicolegal, and society at large. The questions raised throughout the first two chapters are encountered on a daily basis in those areas where government and science are expected to come together in a cooperative spirit.

Nevertheless, an appreciation for the ramification of those interactions between government and the scientific communities is essential in today's world. Science and government constantly interact in a variety of ways, one serving the other and both in service to American society.[2] In the next section the focus will turn to interactions arising through legal constraints on pharmaceutical innovation and on products of recombinant DNA as potential environmental hazards. But first some exploration into the nature of likely interactions is in order.

Government and Innovation

> "Science, and the broad and long term public welfare, have little to do with the way in which economic and societal interests are balanced in the structure of government and in its programs and services."[3]

First and foremost, the federal government encourages and supports the scientific endeavor. The starting point is the Constitution itself, which calls upon the Congress to "promote the progress of science and useful arts."[4] Just how far we have gone in that promotion and how science has filled our lives would certainly be a source of astonishment for the Founding Fathers.[5]

While innovation remains the province of the Patent Office, technology and the sciences, both "pure" and "applied," are impelled in large part by the federal bureaucracy. Science, after all, cannot determine its place in human experience as it evolves. To make that determination, it is necessary to go beyond science to envision the course of history, an objective that may be unattainable.[6] Science policy[7] as such is beyond the scope of this effort. The focus here will necessarily be considerably sharpened to a small component of the larger panorama.

Certainly, basic research is essential to the two agencies we will be addressing in depth: the Food and Drug Administration (FDA) and the Environmental Protection Agency (EPA). But, just as attention has been diverted from basic to applied science and technology in relevant bureaucracies, attention here will be directed to the regulatory utilization of technical information in controlling innovation and its applications. At that juncture, public participation is most likely to have a significant influence on the proceedings. Ironically, the point at which the natural and social sciences are meeting is also the point at which natural science is least developed and least reliable.[8]

In a sense, government impells scientific and technological advancement both generally and for specific purposes. Among those purposes, human health and welfare are paramount. Pharmaceutical research and development serve human health directly. Advances in this and other fields are, however, not without their attendant problems. At that juncture we turn to that uneasy marriage[9] uniting law and science for protection, often to limit or to redirect ongoing advancement.

All the sciences have a role in the unfolding developments. The natural sciences are the inescapable foundation. Technology turns certain fundamental principles to that progress of science and useful arts encouraged under our Constitution. The social and political sciences determine how that progress is to proceed, to what ends and subject to what constraints. Often in our contemporary politicolegal system these constraints take the form of administrative regulations and standards. These can be promulgated only through additional interaction with the relevant sciences and technologies.[10] Finally, the constraints themselves spark the imagination of scientists and engineers. Innovations follow, which interact with the social and political sciences, and new or revised regulatory responses with a technical base lag not far behind.[11]

That science policy itself is not high on the priority list of the electorate is a frequent source of commentary. During the 1988 presidential campaign, a spokesman for the scientific community asserted that politicians had for too long been able to gain office without being able to take an informed position on matters of significance to scientists or engineers and, more to the point perhaps, to society itself.[12] Alvin W. Trivelpiece of the American Association for the Advancement of Science further asserted that support of vigorous research and development is one of only a few ways in which Congress can promote investments likely to be of enduring benefit to the present and to future generations.

Upon election President-elect Bush promised to upgrade the office of science advisor and to ensure access to the presidency and to appoint a committee on science and technology.[13] According to Daniel E. Koshland, Jr., the qualifications for science advisor include wide contacts throughout the scientific community and the capacity for listening coupled with advocacy skills. The position calls for a vision of the future and intellectual leadership on all the diverse problems science must address, not the least of which is to advise the president on his numerous scientific appointments, among them those to EPA.[14] However one may judge a chief executive's actual performance, the attention on specific issues of science during the 1988 campaign is instructive and promising, at least as a first step at the national level.

Innovation through research and development is of vast importance to the future of this nation. But the endeavors of scientists and engineers are also subject to governmental scrutiny, if not to regulation comparable to that constraining other endeavors. There can be no doubt that innovation is to be judged and evaluated both internally by the practitioner's peers and externally by agencies of government.[15]

A question understandably arises: Are those who judge free of conflict of interest? Are those who are judged evaluated by their peers? The expected affirmative answers may not be as facile as might be suspected, and the confounding circumstances have implications for science and technology in the regulatory system. In fact, the two principles of evaluation may well be in conflict, as is revealed in such queries as whether a biologist would agree that the nuclear industry can be self-regulating or whether a member of Congress would agree that the National Institutes of Health can be expected to evaluate biologists who have received grants from that agency. On the other hand, laughter is the unfortunate but likely response of a nuclear physicist who is asked whether Congress can regulate itself.[16]

In short, while modern specialization in science and technology renders it inevitable that evaluation depend on relevant expertise as a matter of basic fairness, each discipline can be expected to suspect that practitioners in another profession are "thick as thieves and only outside observers with no axe to grind can protect the public's interest."[17] No doubt, most members of the public not only would be in complete agreement but also would harbor suspicions of the represented community and of fellow members of society as well. Still, somehow, the impression is that the whole thing works

> in part because the outside world is always watching. The inside group is needed to provide detailed rules and sophisticated analysis. But inevitably, the big picture can be explained to outsiders, and outside groups do intervene if the insiders' decisions seem unwarranted. Almost invariably, sloppiness or negligence in designing procedures develops into a major scandal in which the outsiders demand reform.[18]

Obviously, reasonable compromise between expertise and conflict of interest must be reached again and again. Common sense should prevail. Nevertheless, the fact of the matter is that, as one example, firms that deal with hazardous wastes have to consult on the management of those wastes, just as scientists must evaluate scientific proposals. To recall history's lesson that either excessive suspicion or unthinking neglect can destroy a system is of the essence.[19]

As has been related in chapter 1, public advocacy groups are something of a reflection of both suspicion and neglect. Their existence is a warning that citizen movements must participate in the relationship between government and innovation.[20] Such movements, whether grass roots or a formal national organization dedicated to a specific cause, arise in the very context of conflicting ideas of legitimacy in which policy formulation and innovation are interacting. Unfortunately, demands on time and for technical competence render effective participation by the ordinary citizen increasingly difficult, hence the delegation of roles to trained experts. Expert judgments are relied upon as legitimate because they are perceived to be based on professional, rather than political or personal, considerations.[21]

Nevertheless, those who render policy decisions must remain accountable to the public. It must never be forgotten that government actions are legitimate only when based on the consent of the governed. One expression of consent is active participation in those actions.[22] A requisite of participation, too often overlooked, is a basic understanding of the bases for action.

One major thrust of government and innovation is social regulation. How are these programs to be evaluated? Where can they go wrong internally, before the external pressures of public interest are taken into consideration? Before moving on to more specific examples, it is helpful to consider public health and the environment in the larger context of government and innovation.

Human Health and Safety

> "[R]egulatory experiences . . . indicate that it is almost inconceivable for a marginal scientific school [of thought] to dominate the entire spectrum of decisions. . . . The primary concern . . . is not how to guard against capture by science but how to harness the . . . expertise of the scientific community . . . to advance the public interest."[23]

Some of the answers to the questions raised above are found in an evaluation of medical care for the elderly.[24] The recommended starting point is consistency in the focus of the program at several points: in the conception and identification of the problem, in corresponding policy formulation, and in the development and implementation of regulatory action. The pertinent questions are, whether the general policy is consonant with the parameters of the problem and, then, whether there is a fit between the policy and its program.[25]

An affirmative response to the first question requires adequate conception of the nature and root causes of the problem and an accurate assessment of its scope and dimensions. In public health and the environment the fundamental sources are scientific. The science of the issues must then be blended with technological, economic, social and, ultimately, political considerations.[26] Determining the extent of the fit between policy and its implementation requires evaluation of the basic procedures and the rules themselves.[27]

A mismatch can occur at the point of definition of the program policy or at the juncture where the program is legislated. Sources of a mismatch include lack of appropriate data, lack of resolution among opposing philosophies, or failure of political compromise among vested interest groups.[28] In reviewing the situation one must define the problem, recognize policy as a set of generalized goals and objectives (both express and implicit), see the implementing program as a formalization of the resultant course of action, and recognize the array of confounding factors. Among the latter are multiple perspectives, availability of information, competing social values, and alternative definitions. It is crucial to keep in mind, moreover, that social values are dynamic, that compromise follows from competing philosophies, and that implementing programs alter.[29] That dynamic gives rise in part to perceptions within the regulated communities of perpetual adjustments in compliance aiming to meet regulatory targets in constant motion.

Because science and technology have become indispensable to sound regulation in the areas of interest here, agencies have established science advisory panels. The ostensible purpose is to free policy decisions from politics.[30] The argument is raised, however, that such independence is untenable. The resultant decisions have been subject to challenge both judicial and political, with the quality and sufficiency of those very technical decisions being assailed.[31] This result can be attributed to the basic natures of law, science, and society.

Though science may be beyond the realm of both partners in policy formulation, tradition has it that, while law generally follows society as the latter changes, law sometimes leaps ahead. Gaps in communication are both a cause and a result.[32] Society can perceive legal standards as alien. There is, then, a three-way gap in perceptions where meaningful integration is the order of the day. The gaps are widened as additional disciplines enter the fray.

In policy formulation a mixture of preferences, assumptions, and values is obviously present. What is less obvious is the exact nature of the mixture in a given situation. Moreover, the preferences of the majority may not be effected. In fact, a minority that is close knit, focused, and aggressive can often prevail, especially when the majority views are either indifferent or underinformed. The whole dilemma is exacerbated because public officials can become skillful in professing one policy while actually effecting another.[33] Moreover, the scientist and engineer retained by administrative agencies are themselves human and thus "political animals."

Sheila Jasanoff argues from a number of actual situations in the Food and Drug Administration and in EPA that science in a policy setting is always colored by values, inherent to both science and scientists and to human nature itself. How, she asks, can these experts perform with impartiality? Her answer, in addition to statutory safeguards such as public access to information, is the establishment of clearly identified boundaries separating the underlying science from policy.[34] When such boundaries are established and remain steadfast, both the regulators themselves and, more significantly to this discussion, the public can be expected to accept the expert's contribution as controlling. Then the corresponding recommendations of the scientific panel is invested with "unshakable authority."[35]

A national goal must be to eliminate some of the mystery of science and technology in order to reenfranchise those whose scientific illiteracy has relegated them more to the role of ignorant intruder than productive participant.

Environmental Protection

> "Before the Earth . . . can become an object of human respect and concern it must first become an object of awareness and understanding."[36]

For all its novel components and relative sophistication, the current environmental movement emerges from the roots of historical tradition, including notions of the "simple life."[37] In the nineteenth and early twentieth centuries environmental policy was debated in terms of preservation and conservation.[38] Conservation, at least, reflected public reaction against pervasive environmental destruction, which was attributed to an unrestrained marketplace in natural resources.[39]

Whatever the tradition, it is essential to bear in mind that our planet is in effect two: one a creation of the cosmos and quite independent of humankind, the other a conceptual creation of human information and experience.[40] As Lynton Caldwell suggests, we strive today to maintain the congruence between those two planets.

Today's environmental movement is a part of that effort. In the context of the cosmic and empirical earths we acknowledge the complexity of merging environmental programs with both public policy and science. An understanding of formulated policy starts with comprehension of the environmental movements, with all their own contradictions, and an appreciation of the role of science in shaping societal perceptions of the human relationship with the planet.[41] Of all the policy choices intended to reconcile us with our environment, the most promising is certainly to "enlarge, refine, and extend public understanding of environmental issues and their significance."[42]

Scientists are similarly calling for participation on the part of the scientific community.[43] As the publisher of *Science* puts it, there was no "press" on the "science vote in Iowa" during the 1988 election campaigns, because candidates were not asked how they would address such problems as greenhouse gases and global warming or innovations in genetic engineering. It remains possible to be elected without being informed on such critical issues of science and technology. Scientists, according to Alvin W. Trivelpiece, must alter that situation through participation in the political process. Significantly, with that participation comes a responsibility. In the case of scientists that responsibility is to avoid allowing their own political biases to lead them into misleading candidates on factual matters.[44] Resolution of these and other critical problems depends upon scientific resolution, at least by way of foundation, and science must provide intellectual leadership.[45]

There is a sense of urgency to the call for integration of law, science, technology, and society. As that integration is sought in that arena of environmental protection that is the politicolegal system, certain points cannot be reiterated too often:

> The interpretation of events in the environment may be as much a matter of economic and political interests as of the environmental facts *per se*. . . . Not all environmental events are amenable to unequivocal scientific explanation and evaluation. Some are, however, and to the extent that these explanations are accepted as real and important they may add incrementally to the mental images that people hold regarding the way the world works.[46]

Nothing is new in either the call for scientific participation or the difficulties attending integration of disciplines and perspectives in environmental matters, as one example of this phase of governmental relationships with innovation. Arguably, the contemporary environmental movement formally commenced with the enactment in 1969 of the National Environmental Policy Act (NEPA),[47] to be discussed in greater detail in the next section. The commentary arising from this era of environmental policy formulation is massive, but a few selections can speak for the bulk of the writings while revealing the extent of relevant progress.

Recognition came early that the core of basic information for establishing pollution control standards, which followed hard on the heels of NEPA, was minimal, if not lacking.[48] That environmental protection is ultimately a matter of social policy and the associated politics was not overlooked.[49] Professor Wengert was prophetic when he observed that environmental policy formulation would be made under conditions of extreme uncertainty and that the choices made would be difficult and hazardous. Much is decided with little indication of any public consensus on desired levels of improvement in environmental quality.[50]

Once evolving pollution control programs were well on their course, that the programs and underlying technical definitions were political was not lost on the commentators.[51] Concern about public opinion was expressed with a projection of uncertainties that loom large today. There were no data to support any correlation between public opinion and governmental efforts at pollution control. Even were data available, any perception of cause (public opinion) and effect (government action) was viewed with suspicion.[52] Suspicion is a major characteristic affecting pollution control decisions. If EPA takes an aggressive stance, it risks perceptions of naivete and a misplaced turn back to nature. If the agency is instead more lenient, it is seen as captive to the power of a regulated community.[53]

By the early eighties commentators were calling for reform of environmental regulation.[54] Yet, as noted above, little apparent attention was paid to these points during the 1988 campaign.[55]

Whatever the future of environmental protection holds, certain realities must be taken into account. Science is the initial source with regard to the existence of an environmental hazard and its causes, but it is up to the government to determine what is to be done.[56] Government programs, at least theoretically, are the manifestation of the public will as to the matter. Complicating the situation are a number of factors.

First, the facts and values of the scientific answer must be sorted out. Then the associated simplifications, qualifications, and conditions must be scrutinized,[57] although they are seldom heeded by the general public. Finally, as indicated in another context above, the scientist must have been asked the right question in the first instance.[58] Of that there is no guarantee at all.

Today the additional call is for integration in the conception of resolutions in face of unprecedented demands. The gaps between natural and social sciences and between the sciences and the humanities must be narrowed, and science has its role in realizing this needed integration.[59] In the environmental area we are called upon to recognize that the endeavor will be a monument expressing national character and aspirations. Among others the integrity of ecosystems will be a symbol by which we will be known and judged.[60] Along the way, scrutiny of the politics of scientific advice in areas of human health and welfare, the latter as represented by environmental programs, opens the door to exploring the more general conflict between democracy and technocracy.[61]

LAW, ART, AND TECHNOLOGIES OF ENVIRONMENTAL PLANNING[62]

> "Economics . . . is much more susceptible to varied interpretations than are the natural sciences [including ecology]. The compatibility of economic and ecological values depends greatly on what kinds of economics and what kind of ecology are being compared. . . . There are no value-free methods for assessing the relative merits of alternative economic or ecological propositions."[63]

The Political Framework

> "It is a basic human characteristic, when threatened by catastrophe, to resort to immediate, often violent, action, whether well considered or not."[64]

As noted, in two places the fine art of integrating law, science, technology, and policy is particularly in demand: in regulation of pharmaceutical innovation by FDA and in environmental protection and pollution control as practiced by the EPA. Before moving on to more specific circumstances within those two areas, a pause for a consideration of the law, art, and technologies of environmental planning will lend perspective.

Environmental planning takes on a variety of forms, issuing ultimately from the Constitution and, in certain states, from specific constitutional provisions.[65] The process proceeds through all levels of government in an illustrative exercise in federalism. Planning ranges from the highly philosophical, emphasizing normative environmental ethics, all the way to the daily grind of practicable implementation of specific regulations by administrative agencies and of environmental compliance with pollution control mandates by regulated facilities. There is not a discipline or profession of twentieth century America that does not have a place somewhere in the constantly evolving programs. Public roles infuse the whole process.[66]

In the political process of environmental planning, the public may be looking to scientists for the shaping of policies, but the political choices are said to be more sociological than scientific. Unfortunately, boundaries from one discipline to the next and between factual input and that which is value-laden are not always as visible as they might be.[67] Furthermore, the factual information provided by scientists can be outweighed in the process, not only by economic, political, or technical considerations, but also by sentimental ones.[68] Sometimes raw emotions hold sway. Self-interest is an obvious component of this volatile mixture. But it is worth contemplating the fact that perceptions of self-interest are susceptible to error, leading to the political success of policy choices that may not be generally advantageous.[69] Such contemplation should be accompanied by the accusation that, although society has been called a compact among the dead, the living, and the unborn, the twentieth century is remarkable in that "[w]e live in the day, forgetful of yesterday and altogether indifferent to the morrow."[70] What is more remarkable is that the accusation came out of the thirties, long before we had entered the era of the regulator as the fourth branch of government and the scientist as a fifth.

The role of science in social regulation at large grew out of a perception that the new "fourth branch of government" was overstating one perspective or another in reaching a predetermined result. From the juncture a new perception arose, that political choices were being hidden behind the even more inscrutable technical expertise.[71] As a result, it has been found that relevant scientific advances do not necessarily correlate with either reductions or increases in conflict over policy. There are, in fact, those who claim that "formal and adversarial . . . regulatory decisionmaking highlights uncertainty, polarizes scientific opinion, and prevents efficient resolution."[72] Moreover, a disconcerting argument can be made that the sociology of science itself, as discussed in chapter 2, interferes with reaching truly objective scientific conclusions.[73] Even if policy and science are clearly distinguished in the course of policy formulation, practice indicates the near-impossibility of eliminating the influence of irrelevant subjective values held by scientists themselves.[74]

National Environmental Policy

> "Much of the success in protecting the environment . . . has . . . been brought about by shifting coalitions of groups pursuing economic, recreational, or aesthetic interests that may on other occasions be in conflict."[75]

With Earth Day 1970, public imagination had turned from the long struggle to cope with nature's challenges to preservation of the natural world. As argued in chapter 2, the latter half of this century marks the third great revolution in human existence. The industrial revolution, the scientific revolution, and profound advances in agriculture related to these two most-recent revolutions demanded a hard look at associated environmental repercussions. The modern environmental movement was taking form.[76]

Obviously, the environmental movement did not escape the pitfalls described here. Indeed, it is an embodiment of them. With a distinct beginning in the late sixties, environmental concerns have represented an exercise in crisis management. Contemporary environmental crises can be traced in major part to efforts to solve problems neglected by our immediate predecessors.[77] In the mid-seventies one observer indicated the societal response to environmental problems as ignorant and naive at best and as callous and belligerent at worst. According to Justin W. Leonard, a research zoologist, we go so far as to neglect acknowledging ourselves as a biological community.[78]

Even in our imperfect society, it is through law that we seek to express our vision of what is acceptable and what is not, what is right and what is wrong.[79] In the law of environmental protection, NEPA can be understood as the first step in articulating that vision.

NEPA is an expression of governmental stewardship in our public heritage, with special reference to our natural resources, our environment. That expression finds detail in a series of environmental programs and in the body of federal pollution control statutes.[80] Here we encounter EPA and its efforts to effect the mission through presumably disinterested experts with special training, to the end of dealing in a detached manner with environmental problems. Significantly, detachment is intended from both political patronage and parochial self-interest.[81] But the result too often has been diversion from the mission in pursuit of procedural compliance, a major complaint with regard to NEPA.[82]

Leonard identified the need for policies that allow the development of objectives in a manner not foreclosing other options. But in the same paragraph he observed that we as a species take our environment for granted, with as little curiosity about our world as an actor devotes to the stage: "He is an actor, not a carpenter."[83] That much remains to be done in this and other areas where science and law are working together to the intended benefit of the public is indicated in the experience of a law professor teaching at the intersection. Harold P. Green, writing in 1990, tells of bafflement among his legal colleagues, both teaching and practicing, when he mentions his academic pursuits.[84]

The relationship between NEPA and innovative technology is a harbinger of environmental regulation of releases related to products of recombinant DNA. Technological advances have increased public concerns regarding risk. Innovation carries hazards for which both probability and magnitude are unknown. A society relatively free of worries about basic economics can devote energy to the costs of progress. Contemporary perceptions of those costs have led to strong resistance to any risk, a resistance leading to further polarizations.[85] It is in this setting that we first turn to pollution control and its technologies.

Technologies of Pollution Control

> "The cutting adrift of science bearing on policy from its traditional moorings in academic and industry laboratories emerges in retrospect as

> a major factor in promoting controversy over the technical underpinnings of regulatory decisions."[86]

A cogent but ironic argument has been asserted to the effect that "machine technology" has given life to novel "intellectual technologies" as an adjunct to the environmentally inspired regulation of the former innovations. Both technologies, the argument concludes, harbor unseen dangers.[87] Indeed, a peculiar pattern has arisen in regulation through the two agencies of interest here. Starting with the proposal of an administrative agency to control a given chemical resulting from internal scientific review, a startling and revealing process ensues of challenges by outside experts, further review of the data, modification of the administrative scientific position, withdrawal of the regulatory initiative, and damage to the agency's scientific and political reputation.[88] No matter the outcome, litigation is almost certain to follow. Judicial intervention is hardly more promising. It has been held to be founded on excessive simplification of both science and policy, although the courts are seen as preferring the conclusions of public review to scientific foundations.[89]

Even when the focus does remain on the scientific foundation, scientific stances with regard to every policy, statute, or regulation will be at odds with one another. The situation calls for integration by all concerned in conceiving resolution of unprecedented demands. Although science can take a role in narrowing and filling gaps between the natural and social sciences and between the natural and social sciences and between those sciences and the humanities, we as a society need to be alert to deficiencies in understanding and seek out a meaningful synthesis of what knowledge is providing.[90]

It is the thesis of this book that we have the means for doing so, if we implement them in keeping with this injunction. But first, we need to have an idea of where we stand at the beginning of the newest "Decade of the Environment."

WHERE DO WE STAND?

> "[T]he committees attached to EPA and FDA do indeed help the agencies define good science—and this consensus view of science in turn influences policy—but they perform this function in part through skilled boundary work and in part through flexible role-playing."[91]

As we enter the nineties, it might be said with some justification that environmental protection and its professions have come of age. Negotiation is beginning to stand on footing as solid as that supporting fervent advocacy. By way of example, the regulatory practices of FDA and EPA are models of negotiation not only among scientists but also between them and the lay public. For good or ill, science alone cannot muster sufficient authority to resolve disputes on the small or the grand scale.[92]

At this juncture commentators are urging what can be described as something of a return to ecological soundness in environmental planning, for application on a scale extending ultimately to the global.[93] At the same time, however, there are cries of foul. Critics of existing law in point claim that the programs are irrational and even vindictive, especially in what amounts to demands for the nonsensical impossibility of a no-risk society. Part of the irrationality is found in the suspicion that, for all the tedious litigation and bureaucratic inertia, many major hazards are going unregulated because attention is diverted to the seemingly trivial.[94] The overall problem is attributed in part to the inability of the law to handle ultimate causes, as opposed to end results and action.[95] Moreover, the pollution control approaches instituted in the seventies and eighties are running out of steam, with new programs expected to be more expensive and complex, with little hope for more than marginal benefits. In short, the easy improvements on behalf of the environment have been made.[96] As we move into the more subtle and more difficult ecological efforts, "[w]e cannot pretend that ecosystem protection can be solved by a large intrusive bureaucracy generating endless regulations while ignoring the population and consumption elements."[97]

The need for well-prepared, liberally educated scientists and other environmental professionals is increasingly apparent, as is the need for an electorate with whom effective communication can be maintained. Observers from government and industry claim that qualified personnel cannot be found.[98] One member of academia asserts a need for a comprehensive program " 'attacking all the problems simultaneously, from training "environmentally literate" scientists and engineers to creating public awareness of the . . . threat to the environment.' "[99]

A glimmer of hope might be seen in the appointment of a science director by Greenpeace U.K. That group, better known to some for fearless action than for facts, seems to be showing signs of recognizing that environmental matters are subtle, complex, and multifaceted. An additional manifestation of that recognition is the Greenpeace Education Trust, which will be funding research that will be subjected to peer review. Further hope for direction in the future is found in the qualifications of the new director. He is a man who combines "academic expertise with the ability to absorb and communicate abstruse information."[100]

Here in the United States it appears that the National Science Foundation is taking action, albeit addressing scientific illiteracy in general rather than in the specific context of human health and welfare as protected through the programs of FDA and EPA.[101] Such programs will be considered in some detail in chapter 11.

In turning now from the general to the specific, it appears that public participation in environmental matters of both local and global concern—

which can also be described, respectively, as personal demands and stewardship—requires enlightenment at two levels, the scientific and the ethical. The goal of scientific enlightenment is not a single, perfect, verifiable truth but

> a serviceable truth: a state of knowledge that satisfies tests of scientific acceptability and supports reasoned decisionmaking, but also assures those exposed to risk that their interests have not been sacrificed on the altar of an impossible scientific certainty.[102]

> At the ethical level, we must seek out a comprehensive ethic for environmental policy [heretofore] hampered by the circumscribed ethical positions of tightly focused disciplines and organizations. Attempts to broaden both ethical positions and the training of individuals for environmental missions are handicapped by sectarian thinking and by entrenched interests, both academic and administrative, that see in synthesis a threat to their security.[103]

Only if we successfully integrate these two lines of endeavor, along with the major part of their impedimenta, can we hope to prevent the demise of law embracing postulates of science and technology as predicted by Aristotle.[104] Nowhere is the integration more urgent than in FDA and EPA.

NOTES

1. Koshland, *Science Advice to the President,* 242 SCI. 1489, December 16, 1988, editorial. Daniel E. Koshland, Jr., is the editor of SCIENCE.

2. For a detailed examination of these interactions as reflected in the judicial system, *see* Kreiling, *Scientific Evidence: Toward Providing the Lay Trier with the Comprehensible and Reliable Evidence Necessary to Meet the Goals of the Rules of Evidence*, 32 ARIZ. L. REV. (4) 915 (1990).

3. L. CALDWELL, BETWEEN TWO WORLDS: SCIENCE, THE ENVIRONMENTAL MOVEMENT, AND POLICY CHOICE (1990), at 81. Lynton Caldwell is at Indiana University in the Department of Political Science and the School of Public and Environmental Affairs.

4. U.S. Constitution, Article I, Section 8.

5. Indeed, the extent of the bureaucracy itself would come as something of an unpleasant surprise. *See* J. BECK, OUR WONDERLAND OR BUREAUCRACY: A STUDY OF THE GROWTH OF BUREAUCRACY IN THE FEDERAL GOVERNMENT, AND ITS DESTRUCTIVE EFFECT UPON THE CONSTITUTION (1932). James M. Beck was Solicitor General in the Hoover administration. This earlier volume makes an interesting comparison to that of Senator Douglas. *See* chapter 1.

6. Caldwell, *supra* note 3, at 5. For an argument to the effect that society experiences a profound impotence with regard to channeling innovation, *see* TRIBE, CHANNELING TECHNOLOGY THROUGH LAW (1973), at 1. Professor Tribe warns, however, against extremes: either making technology the scapegoat for all

societal ills or accepting it for its own sake as opposed to being a tool for society. Ibid. at 2. (For the view in the mid-seventies, *see* note 51, *infra.*) Compare this situation to its counterpart in chapter 2; science itself is impervious to social pressures.

7. Appropriately in the context of this book, the pervasive demands government and the technological world impose on each other are suggested in an array of sources, ranging from the philosophical through the scholarly (notably including Tribe, ibid.) to the frankly popular. The following publications are no more than a sampling. Davis and Rogers, *Dukakis, Bush Duel on Science-Policy Promises as Scientists Fret over Funding, Focus on Research*, WALL ST. J., August 9, 1988, at 52; NATIONAL ACAD. SCI., THE SCIENTIFIC ENDEAVOR: CENTENNIAL CELEBRATION OF THE NATIONAL ACADEMY OF SCIENCES (1963); Shapiro, *Fragmenting and Reassembling the World: Of Flying Squirrels, Augmented Persons, and Other Monsters*, 51 OHIO ST. L.J. (2) 331 (1990).

8. Caldwell, *supra* note 3, at 67. *See also* T. HOBAN and R. BROOKS, GREEN JUSTICE: THE ENVIRONMENT AND THE COURTS (1987). Thomas More Hoban and Richard Oliver Brooks visualize administrative agencies as a fourth branch of government. Uniquely, agencies possess two characteristics, longevity and expertise, which render them capable of enormous influence on our lives through utilization of scientific and technical expertise in combinations found nowhere else. Hoban and Brooks, at 57–8. Ironically, those same characteristics also give rise to the many drawbacks of which the bureaucracy is subject.

9. The metaphor is hardly novel and can come as no surprise. *See*, e.g., Markey, *Science and Law: Toward a Happier Marriage*, 59 J. PAT. OFF. SOC'Y (6) 343 (1977); *Science and Law: A Dialogue of Understanding*, 68 A.B.A. J. 154 (1982); Davis, *The 'Shotgun Wedding' of Science and Law: Risk Assessment and Judicial Review*, 10 COLUM. J. ENVTL. L. 67 (1985); Wessel, *Adversary Science and Adversary Scientist: Threats to Responsible Dispute Resolution*, 28 JURIMETRICS J. 379 (1988). *See also* chapter 9.

10. *See generally* S. JASANOFF, THE FIFTH BRANCH; SCIENCE ADVISORS AS POLICYMAKERS (1990). In 1986 Sheila Jasanoff was the director of Cornell University's Environmental Law and Policy Program

11. For a critique of this process in the environmental protection/pollution control area *see* Brooks, *Intellectual Technology: The Dilemma of Environmental Law*, 15 RUTGERS COMPUTER AND TECH. L.J. (2) 411 (1989).

12. Trivelpiece, *The Science Vote in Iowa*, 241 SCI. 513, July 29, 1988, editorial. Alvin W. Trivelpiece wrote as the publisher of SCIENCE.

13. Koshland, *supra* note 1.

14. Ibid. The place of science policy among the media and, by inference, in society at large is suggested by the somewhat forlorn observation that the resignation of the science advisor "might" share headlines with the announcement of a new Redskins' quarterback.

President Bush named D. Allen Bromley his science advisor. It was Bromley who announced the administration's 1991 budget which emphasized "big science." Included was a 7 percent increase in spending for research and development. Those programs recognized through the budgetary process are the space program, the National Science Foundation, and the Department of Agriculture. Reviews from the scientific community were expected to be mixed, especially since congressional approval was deemed unlikely. Notably, however, funding for biomedical research was increased beyond inflation—an unusual move, but one representing the Human Genome Project on one level and pressures for AIDS research on another. In the aggregate, research on global environmental changes received a 57 percent increase over that enacted for 1990. On the other hand, programs including the National Oceanic and Atmospheric Administration's research were subjects of less financial recognition. Norman, *Bush Budget Highlights R&D*, 247 SCI. 517, February 2, 1990.

15. Koshland, *Conflict of Interest*, 249 SCI. 109, July 13, 1990 editorial.

16. Ibid.

17. Ibid.

18. Ibid.

19. Ibid.

20. *See*, Goslant, *Citizen Participation and Administrative Discretion in the Cleanup of Narragansett Bay*, 12 HARV. ENVTL. L. REV. (2) 521 (1988), at 521. Kim Herman Goslant was a J.D. candidate at Harvard at the time of writing.

21. Ibid.

22. Ibid. Mr. Goslant asserts that there is another conflict in determining consent, in that individuals paradoxically seek an active role in community affairs that may run counter to choices made in voluntary consumer transactions. Ibid., at 522–3. This is a paradox unlikely to dissipate in the foreseeable future. For more on the public roles in a similar environmental context, *see* Mansfield, *The "Public" in Public Land Appeals: A Case Study in "Reformed" Administrative Law and Proposal for Orderly Participation*, 12 HARV. ENVTL. L. REV. (2) 465 (1988).

23. Jasanoff, *supra* note 10, at 250.

24. H. BREHM and R. COE, MEDICAL CARE FOR THE AGED: FROM SOCIAL PROBLEM TO FEDERAL PROGRAM (1980).

25. Ibid., at vii.

26. *See*, e.g., Jasanoff, *supra* note 10, *passim.*

27. Brehm and Coe, *supra* note 24, at vii.

28. Ibid.

29. Ibid., at viii. For example, if a definition of the problem is in error, the result can be policy intentions wide of the mark. In this case there is little hope of resolving the problem. Ibid., at 41. But the very complexity of problems for which the federal government seeks resolution is likely to preclude the simple definitions readily understandable to those without relevant expertise. Ibid. Care must be taken to avoid outdated scientific studies (ibid., at 52), as well as those lacking validity among its proponents' spheres, although a case can be made for the proposition that pseudoscience or quackery will not prevail today. *See* Jasanoff, *supra* note 10, at 250.

30. Jasanoff, *supra* note 10, at 1.

31. Ibid.

32. Brady, *"But Most of It Belongs to Those Yet to be Born": The Public Trust Doctrine, NEPA, and the Stewardship Ethic*, 17 B. C. ENVTL. AFFS. L. REV. (3) 621 (1990), at 630–631. Timothy Patrick Brady is in good company. Aristotle long ago observed that if law in the aggregate conflicts with the spirit (the genius or *ethos*) of the people, the certain result of the conflict will be the demise of the conflicting law. From a quotation in Beck, *supra* note 5, at x.

33. Caldwell, *supra* note 3, at 77.

34. Jasanoff, *supra* note 10, *passim* and at 5, 39, 236–237.

35. Ibid., at 236. This commentator would urge agencies and their respective advisors to work together in establishing those boundaries. Notably, according to

Jasanoff, along with most aspects of administrative policy formulation, the boundaries are a matter of negotiation. Arguably, one means of fostering meaningful citizen participation is to make those boundaries express whenever the bases of regulation are made accessible to the public.

36. Caldwell, *supra* note 3, at 38.

37. *See*, e.g., R. NASH, WILDERNESS AND THE AMERICAN MIND (rev. ed. 1967), and D. SHI, THE SIMPLE LIFE: PLAIN LIVING AND HIGH THINKING IN AMERICAN CULTURE (1985).

38. Those traditions are alive and well, their vitality ensured through a number of statutes of current interest. *See* WEST PUBLISHING CO., SELECTED ENVIRONMENTAL LAW STATUTES (Educational ed. 1990–91). The contents include representatives from U.S. Code Title 16 (Conservation), Title 33 (Navigation and Navigable Waters), Title 43 (Public Lands)—all of which reveal a heritage in preservation as well as in conservation.

39. Goslant, *supra* note 20, at 524.

40. Caldwell, *supra* note 3, at 39.

41. Ibid., at xi.

42. Ibid., at xiii. Two others, reflecting diametrically opposed perspectives, are popular: To accept political and economic realities and seek mediation and compromise or to engage in guerrilla warfare against ecologically destructive practices. Ibid., at xi–xiii.

43. *See*, e.g., Trivelpiece, *supra* note 12.

44. Ibid.

45. Koshland, *supra* note 1.

46. Caldwell, *supra* note 3, at 76.

47. 42 U.S.C.A. 4321–4370b.

48. *See*, e.g., E. DERAN, POLLUTION CONTROL: PERSPECTIVES ON THE GOVERNMENT ROLE (1971). This volume was published by the Tax Foundation, Inc., a private, nonprofit organization for nonpartisan research and public education on fiscal and management aspects of government.

49. *See*, e.g., N. Wengert, *Environmental Policy and Political Decisions: The Reconciliation of Facts, Values, and Interests*, in POLITICS AND ECOLOGY (P. Foss, ed., 1972), 39. Norman Wengert wrote as a professor of political science at Colorado State University.

50. Ibid., at 40–42.

51. *See*, e.g., C. DAVIES and B. DAVIES, POLITICS OF POLLUTION (2d ed., 1975). These authors recognize the place of public education in this arena, but they also see what might be characterized as a perversion of the political system as manifested in the rampant growth of bureaucracy. (*See also* Beck, *supra* note 5.) Successful education can result in pressure on Congress for more support of the responsible agency (such as EPA), which in turn leads to greater activity leading to a need for more education–with potentially unlimited extension. Ibid., at 85.

52. Ibid., at 87.

53. Jasanoff, *supra* note 10, at 23.

54. *See*, e.g., REFORM OF ENVIRONMENTAL REGULATION (W. Magat, ed., 1982). Wesley A. Magat was associated with the Center for Study of Business Regulation, Fuqua School of Business, at Duke University. His book provides par-

ticularly informative analyses of the politics of environmental protection and pollution control for anyone studying the law, art, or technologies of environmental planning.

For a more recent and more succinct analysis, *see* Sagoff, *Where Ickes Went Right*, or *Reason and Rationality in Environmental Law*, 14 ECOLOGY L.Q. (2) 265 (1987). Mark Sagoff acknowledges four models of environmental law, including one founded in public values. He asserts that, of the four, that based in public values engages in moral deliberation rather than mathematical aggregation. None is proferred as a tool for integrating the array of contributions and interests. Ibid., *passim*.

55. For one popular impression of the situation, *see* Muro, *Quiet: Environmental Politics at Work*, BOSTON GLOBE, May 22, 1988, at 85. According to the article the campaign was "almost bereft of dialogue on the environment," with neither presidential candidate discussing toxic wastes, acid rain, or water pollution. Ibid.

56. Susskind and Weinstein, *Toward a Theory of Environmental Dispute Resolution*, 14 LAND USE & ENVTL. REV. 433 (1980–1981), Note 42 on 446–447, at 447.

57. Ibid. *See also* Jasanoff, *supra* note 10, at 49.

58. Susskind and Weinstein, *supra* note 56, at 447. One problem with environmental questions as ordinarily posed is that anthropogenic alterations in ecosystems are judged on the criterion of human approval. Our evaluations, then, are human and guided less by ecology and more by economics, politics, morality, and aesthetics. *See*, e.g., Hoban and Brooks, *supra* note 8, at 5.

59. Caldwell, *supra* note 3, at 123.

60. Sagoff, *supra* note 54, at 321.

61. Jasanoff, *supra* note 10, at 2.

62. This heading is the subtitle of an interdisciplinary seminar the author has taught to a mixture of practicing professionals, J.D. candidates, and students in a master's program in environmental law and policy at Vermont Law School's Environmental Law Center. The Center and its programs strive for meaningful integration of perspectives for the resolution of environmental problems through a variety of approaches, only one of which is pollution control.

63. Caldwell, *supra* note 3, at 83.

64. Leonard, *Ethics and Realities*, in ENVIRONMENTAL QUALITY AND SOCIETY (R. Tybout, ed., 1975).

65. Constitutional provisions in California and Idaho expressly address riparian rights. The constitutions of Missouri, Oklahoma, Texas, and Washington take at least minimal steps, such as establishing an agency for natural resources, toward recognition of environmental values. Pennsylvania's constitution provides for conservation bonds, that of Nebraska recognizes water as a public resource, Arkansas seeks to conserve wildlife. The constitutions of Alaska and Rhode Island reveal something of a development perspective. On the other hand, Virginia and New York take a conservation approach. States with express constitutional protection of the environment are Montana, Louisiana, Massachusetts, Florida, and Hawaii. The remaining states maintained a constitutional silence with regard to environment and natural resources. D. Howell, Interdisciplinary Seminar/Environmental Planning: Art, Law and Technologies (1990), at 11–18 (unpublished seminar manual).

66. Ibid., *passim*.

67. Caldwell, *supra* note 3, at 13 and 21.

68. Ibid., at 76.

69. Ibid., at 82.

70. Beck, *supra* note 5, at 250.

71. Jasanoff, *supra* note 10, at 3.

72. Ibid., at 8.

73. Ibid., at 12–14, *passim.*

74. Ibid., at 230.

75. Goslant, *supra* note 20, at 548. Mr. Goslant is an articulate advocate of direct and active citizen participation in environmental processes, with the successful involvement of one particular group in the restoration of a specific natural resource. Notably, he calls for a major educational—as opposed to propaganda—component in such participation by such a public interest group. Ibid., *passim.*

76. For a brief outline of the environmental program as it evolved between Earth Day 1970 and 1990 and how law, science, technology, and society interact, *see* Howell, *Intersection/Environment*, 3 NASTS NEWSLETTER (2) 4 (1990).

77. Brady, *supra* note 32, at 622.

78. Leonard, *supra* note 64, at 49.

79. Ibid., at 629–30.

80. *See* note 38, *supra*, for an indication of the scope of these programs, which do not always relate clearly to the ultimate mission of environmental protection as the promotion of harmony between humankind and their environment. *See* 42 U.S.C.A. 4321.

81. Goslant, *supra* note 20, at 525.

82. Brady, *supra* note 32, at 637.

83. Leonard, *supra* note 64, at 56.

84. Green, *The Law–Science Interface in Public Policy Decisionmaking*, 51 OHIO ST. L.J. (2) 375 (1990). Professor Green "seeks to determine how the various processes of law . . . respond to changes brought about by scientific advances." He is not engaged in analyzing computers and other innovations in aid of legal practice, in intellectual property, medical practice or scientific evidence as a component of evidence law. Ibid., at 375.

85. Binder, *NEPA, NIMBYs and New Technology*, 25 L & W L. REV. (1) 11 (1990), at 15.

86. Jasanoff, *supra* note 10, at 39.

87. Brooks, *supra* note 11. For brief discussions of technologies in the milieu of air pollution control and of oil spills, respectively, *see* Howell, *Intersection/Environment*, 3 NASTS NEWSLETTER (4) 5 and (6) 5 (1990). For an insider's view of the role of science at EPA in the nineties, *see Counting on Science at EPA*, 249 SCI. 616, Aug. 10, 1990. Significantly, this feature includes a comparison of EPA's top eleven concerns in contrast to the public's ranking of concerns. Needless to say, the two do not match. *Counting on Science*, at 616. *See also* chapter 9.

88. Jasanoff, *supra* note 10, at 20.

89. Ibid., at 49.

90. Caldwell, *supra* note 3, at 13 and 123.

91. Jasanoff, *supra* note 10, at 236–237. The professional capacities in which these scientists act include peer reviewer and educator as well as technical consultant, mediator, judge and, in the end, policy advocate. Ibid., at 237.

92. Ibid., at 234.

93. *See*, e.g., Sagoff, *supra* note 44, at 267; Brady, *supra* note 24 *passim*, and *$1.5*

Billion Targeted for Global Environmental Effort, January/February ENV'T TODAY 50 (1991).

94. *See*, e.g., Sagoff, *supra* note 54, at 295–297.

95. *See*, e.g., Hoban and Brooks, *supra* note 8, at 55.

96. *See*, e.g., Reitze, *Environmental Policy: It is Time for a New Beginning*, 14 COLUM. J. ENVTL. L. (1) 111 (1989), at 112.

97. Ibid., at 155.

98. *See*, e.g., Kelly, *The Booming (But Begging) Market for Environmental Professionals*, January/February ENV'T TODAY 27 (1990).

99. Ibid., at 28, quoting Kurt Fischer, associate director for research in hazardous waste management program at Tufts University. Help, this article asserts, is on the way in the form of an environmental education bill then pending in the Senate. Ibid., at 29. The 1990 U.S.C.A. at Title 20 (Education) makes no reference to a current provision. In fact, while Chapter 35 provided appropriations for environmental education over the years 1971–1977, those provisions were superseded and then repealed in 1981. 20 U.S.C.A. Chapter 35 (1990).

100. Cherfas, *Greenpeace and Science: Oil and Water?* 247 SCI. 1288, Mar. 16, 1990.

101. Bloch, *Education and Human Resources at the National Science Foundation*, 840 SCI. 839, August 24, 1990 Perspective.

102. Jasanoff, *supra* note 10, at 250.

103. Leonard, *supra* note 64, at 57–58.

104. *See* note 32, *supra*.

PART II

COLLISIONS AND IMPLICATIONS FOR SOCIETAL SCIENTIFIC ILLITERACY

CHAPTER 4

Revolution in Pharmaceutical Innovation and Evolving Policy

PRELIMINARIES

> "The individual fantasies of patients and physicians about sickness and health, . . . and the willingness to bend or suspend rational judgment have remained constant . . . ,but the scale and structure of medicine and science have changed radically in this century."[1]

In the preceding chapter a litany of needs associated with the intersection of science and the bureaucracy was proposed. Among these needs, five stand out as representative of the problems intruding on meaningful public participation. For purposes of discussion, "meaningful" is defined ambitiously. Meaningful participation in policy formulation is based at least in substantial part on recognition of the relevant issues and comprehension of the factual bases for their resolution, reasonably accommodating all the interests, whether presently vested or implicit to the philosophy driving the mission, subject to evolving policy.

With regard to the scientific component of regulatory policy, five requisites can be identified: (1) basic understanding of the fundamentals supporting action, (2) peer review, (3) avoidance of conflicts of interest, (4) public attention, and (5) both trust and trustworthiness. Even when the argument can be made that all five have been met, problems can still arise at the intersection, especially with regard to public participation in the various processes. Here, three major sources of those seemingly inescapable problems are identified: (1) lack of appropriate data in the first instance, (2) failure of political compromise at one level, and (3) lack of resolution among philosophies at another level. Public participation is a major component in the resolution of the three problems. The irony is that the public can also be an unavoidable impediment to meeting the five needs.

This Gordian knot of needs and problems gives rise to a transcending uncertainty haunting the regulatory processes: Is general policy consonant with the problem it addresses and, if so, is there real concert between the policy and its program? The answers in the contexts of pharmaceutical innovation and regulation of genetically engineered products depend on effective integration of the relevant disciplines, clear-cut boundaries between science and policy, and public comprehension of the results of these two efforts. The last component calls for real education and dissemination of information, as opposed to the more usual propaganda and outright brainwashing. If there is an overall philosophical mission here, it is the elimination of mysteries surrounding science and technology in an effort to reenfranchise the American public.

Here and in the following chapters, we will explore first pharmaceutical innovation and then releases from production and use of products created through the novel biotechnology, direct anthropogenic recombinations of DNA, the molecule of heredity. Before us are are the questions, problems, and suggested answers noted above.

As public participation in the control of pharmaceutical innovation and then of environmental protection is addressed in detail, an interesting series of analogies is worthy of thought. It is fitting that thinking along these lines start with the individual, move to one of the deepest of individual concerns, and then move on to concerns somewhat more removed from immediate individual experience. The concept in point is human response to trauma.

Trauma can be experienced at a variety of levels. These levels can be expressed through a series of conceptual partners: social and personal, external and internal, public and private and, separately, sudden and enduring. There are, in fact, what appear to be universal responses to trauma, whatever the level of its experience. The initial response is total denial, followed eventually by what is considered mature reintegration.[2] Between the two extremes lies distortion followed by the process of restoring stability. Normally, the initial sense of total vulnerability is deliberately kept at bay, because the random dangers of life can be intensely frightening, because they are mostly improbable, and because, after all, they are impossible to avoid in any event.[3]

During the early stages of this universal response, one may achieve some momentary satisfaction in fantasizing revenge. The victim may seek out causative agents, often totally unrelated to the actual cause. Thereby the sense of passive helplessness is lessened. More precisely, the experience can be set out in a whole series of steps, which are deemed healthy by professional counsellors: denial, blame, self-blame, deep sorrow, resignation, and acceptance. The ultimate stage and the healthiest is that marked by an emergent sense of meaning and purpose founded in accepting the reality of the trauma. What is unhealthy is to become lost, as many do, in one of those intermediate stages.[4]

Because trauma can occur at a variety of levels and the pattern of response is universal, public responses to the traumata of perceived misdirection in the regulation of pharmaceutical innovation and environmental protection can be examined on the same progression. In examining societal traumata to which regulatory programs are directed, the individual is the indispensible starting point. It is the individual, after all, who is to be served and who is the participant, whatever the capacity.

Relevant law in this country puts the individual first, although always in balance with the rights of the public at large. At both the individual and societal levels, moreover, health comes first, followed in priority by welfare. Here the closest human concern, personal health, is reflected in perceptions attending access to pharmaceutical innovation. The more removed welfare is represented by environmental protection. Recombinant DNA blends the two with the added elements of morality and ethics.

Clearly, in both programs identification of the proper source of the particular societal trauma in question and an emergent sense of regulatory purpose and meaning, based in reality, are the "healthy" public goals. But, when the public faces such a societal trauma with an unhealthy cessation in progressing to that emergent sense, the consequences for the emerging policy can be dire.

Because application of this analogy is instructive, we will be exploring relevant regulation with the placement of political response in the indicated line of progress. The expected effect on policy formulation, if the effective public voice is constrained by distortion and a compulsion for revenge or by resignation, can be disastrous. Similarly, substantial damage to sound regulation can be predicted when major voices among the public are at different stages of the progression. As with public indifference, either situation is bound to be counterproductive at the very least.

THE SETTING OF THE INTERSECTION

> "[F]ar and away the biggest risks—and the hardest to appraise—are the risks of research. Every new drug is intended to make an old one obsolete, and the turnover is accelerating to the point of frenzy."[5]

Revolution in the Science of Medicinals

> "There is . . . an age-old conflict between the desire of the medical profession and the general public for bold new advances and the conservative conviction that the art of medicine works best when it helps nature to take its course."[6]

First, *to do no harm* is the edict of Hippocrates. It remains the guidepost for the medical professions today. Medicinal substances have been with us for thousands of years,[7] but it was not until the last century that knowledge of healing drugs became more than empirical. From the earliest human history therapeutic effects have been observed, but the chemical and biological foundations of those effects remained to be discovered. Even in more recent times, where active ingredients are mixed with inactive ingredients, the one could not be distinguished from the other. As there is no apparent, rational theory of drug effects, the initial discovery of new pharmaceutical agents today remains largely dependent on trial and error and outright serendipity.[8]

The opening of the modern era of pharmaceuticals can be identified in the first half of this century. The year 1932 has been designated as the start of a new era with the development of the antibacterial sulfanilamide, not because of that chemical as such but because genuine science is recognized as the foundation of sulfanilamide's establishment in the pharmacopoeia.[9]

No one can seriously doubt that modern drugs significantly reduce mortality and facilitate recovery. Through pharmaceuticals we can expect to return to our routine from sickness or injury more quickly than ever before and with less residual disability than afflicted those before us. There can be no member of the public who has not had experience with one medicinal or another with substantial benefit, if not personally, then through a family member or friend. The value of therapeutic drugs can be measured in numbers of people or in economic terms.[10] Either way, the value borders on the priceless, especially in the aggregate. But there is an associated hazard.

In the thirties, when the modern pharmacopoeia was just emerging, a group of people died when an improper solvent was used for sulfanilamide. As recently as 1940, properly controlled clinical trials of drugs were almost unknown. Numerous advances were in effect by the sixties. Tests for toxicity in animals and in specific organs were available. A relationship between certain medications and untoward effects on the patient was being taken seriously, but it remained impossible to test for new kinds of toxicity or for effects not previously observed. Science would not reveal whether toxicity in man or animal could be predicted across respective species lines, for all the revelations associated with practices such as double-blind clinical tests.[11] For all our contemporary sophistication, effectiveness (efficacy) itself cannot be predicted across species lines. Drugs that are effective in animals may be without effect in humans and vice versa.[12]

Today we are still faced with the unknown in the development and application of pharmaceuticals. Scientific answers to crucial questions remain indeterminate. Thus, the mission of policy formulation should be to manage relevant scientific and medical issues on the scientific and medical merits in combination with judgment and the incorporation of sound values. Legal merits are ancillary, not primary, and should not be permitted to overwhelm the substance of the matters placed in issue. As is so often the case, benefits must be measured against inextricable risks. At the same time, the public interest must be constantly measured. Often, there is no comprehensive, balanced analysis of what to require of interested sectors or how each requirement should be met. There is no single implicit, obvious societal need. Instead the tendency is to lose perspective as responsible policy evolves in the politicolegal forums.[13]

To do no harm is no simple task in treating patients through the largesse of the modern pharmacopoeia. There are novel features, such as an unprecedented rate of medical advance in the twentieth century with its associated advances in the human stakes put at issue. The development of pharmaceuticals has now entered a political realm wherein scientific facts and medical pronouncements are accompanied by ideological broadsides. The process has become a complicated, multifaceted, and multidisciplinary endeavor where money, reputation, and public honors are put in issue alongside the medicinal values.[14]

Thus, the scale and the structure of medicine, as well as science at large, have changed radically in the twentieth century. Science itself has become an industry. The research establishment, with special reference to the medical arts and sciences, has been raised to an unprecedented scale. Not surprisingly, the social forces unleashed and the ramifications of medical treatment can exceed the power of the individual physician (let alone the individual patient) to control.[15] Pharmaceutical innovation and development, marketing, distribution, and sales are a major part of this recent phenomenon.

The pharmaceutical industry itself assumed its contemporary form during the course of World War II. The stimulus was the development of penicillin and other antibiotics. Research into additional therapeutic fields was comparably stimulated. The rate of new drug introductions increased from a low of less than fifteen per year in the interval of 1941 through 1945 to a peak in the latter half of the fifties of up to 65 per year. The rate remained relatively high through the mid-sixties, but it declined through the seventies.[16]

The human benefits of advances, notably in therapeutic drugs, cannot be denied. It is no longer startling to hear of people who are one hundred years old and counting. Worldwide life expectancy at birth has increased 50 percent since the beginning of the century. Our children are generally heavier, taller, and longer-lived than their predecessors. There are many reasons for these improvements, but significant here are better diagnosis and treatment of our ills and more and better drugs to treat them. Significantly, the prices of these drugs are decreasing over time, and mass production is the reason. Penicillin, for example, cost twenty dollars the dose when first introduced. Two years later a dose cost less than a dollar. Less than ten years from its introduction, penicillin sold wholesale for ten cents a dose.[17]

No one is likely to deny or to begrudge the fact that pharmaceuticals represent a profitable endeavor, but it must also be noted that profits may be derived primarily from only a few products and that research and development is an extraordinarily expensive way of doing business.[18] By 1969 the industry had devoted some forty million dollars to orphan drugs alone, and the federal government invested more than half a billion in drug-related research and development.[19]

Among the costs of research and development are a wide range of risks inextricably and inherently represented by the biochemistry of pharmaceuticals. The significance of these risks is often magnified by public perceptions of hazards. For example, market bonanzas realized by the industry must be balanced against losses associated with public indifference or public antipathy resulting from a well-publicized disaster, even when it is the result of misuse.[20] More recent costs include threats of liability in addition to resources underlying the establishment, preservation, and effectiveness of crucial intellectual properties associated with the industry.[21]

Pharmaceutical research and development is the point at which the U.S. government enters this arena. The federal government plays three major roles here: generator, guarantor, and guardian. The government is instrumental in the discovery and development of drugs, it is their guarantor through the patent system, and it is the guardian of the consumer.[22] The role of public guardian is effected through the Federal Food, Drug, and Cosmetic Act,[23] implemented by the Food and Drug Administration (FDA).

Because the FDA is charged with regulating pharmaceuticals to reduce the biochemical risks posed by drugs, it is important to recognize where those risks may be found. Some risks occur prior to this agency's involvement. The initial risks are those to the industry itself. Any given line of pharmaceutical research may prove to be pure waste. There is no yardstick by which potential return can be routinely and accurately measured. Moreover, each firm is at the mercy of its competitors' research. A product may have a brief lifetime of profitability.[24] Additional risks, normal but unique to this industry, are the facts that the products are perishable, that an extraordinary standard of quality is applied from within and without, and that legal liability is strict, liability absent the element of fault. Finally, research itself presents an ironic risk, that of rendering an existing drug obsolete.[25]

For most responsible individuals, the most appalling risks must certainly be those associated with a drug at work and those unavoidably inherent in the clinical trials needed to establish the actual medicinal value and the hazards of a new drug. While those risks can be reduced before initial marketing commences, they cannot be eliminated. The unexpected will always remain a part of the process.[26] These realities go far in explaining why so high a percentage of sales dollars go into research. One example is revealing: Over an interval when one firm's sales and profits both increased by more than 600 percent, its research and development expenditures increased by more than 2,000 percent.[27]

Though a case can be and has been made for excesses in the promotion of drugs,[28] few of us are likely to be willing to give up the health and welfare pharmaceuticals represent for us. Examples of those perceived benefits as well as perceived excesses abound throughout the recent history of medical therapy and within the lifetime of the FDA. Clinical trials are an essential and highly illustrative component of the processes of scientific advance, medical usage, government regulation, and public perceptions of the whole. Appropriately enough, clinical trials have never before been more expensive, time-consuming, or controversial.[29]

It is probably to our great advantage that we are not bound by a choice between a wholly market-oriented approach to the sensitive issues of pharmaceuticals and one of entirely centralized government control. Because which policy measures will best serve society cannot be predicted, alternative policies must be compared in terms of their respective benefits and costs.[30] That is what the FDA, itself evolving in response to the corresponding scientific revolution, has been seeking to accomplish since its inception.

Evolution of Legal Control of Pharmaceuticals

> "It is probably undesirable for an official body to make detailed regulations for detection of side-effects, since regulations are unlikely to lead to the detection of unexpected effects, and there is a danger that the doctor will think the regulations relieve him of the responsibility of watching for them."[31]

Our system for the regulation of pharmaceuticals has been evolving in this century in response to both scientific advances and alterations in societal expectations of drugs. Major issues encompass not only the safety and effectiveness of drugs in the marketplace but also the quality of the underlying research. Furthermore, the protection of the human subjects of investigational drugs is a major concern.[32]

The Pure Food and Drug Act of 1906 was enacted in response to widespread promotion and unrestricted sales of patent medicines containing opium, heroin, and cocaine absent any indication of the presence of such hazardous substances. Neither labels nor warnings were provided. According to this statute, itself a legal innovation, a drug was adulterated if it failed to meet official standards associated with the name under which it was sold. A drug was misbranded if its labelling was in any way false or misleading.[33] Adulterated and misbranded drugs were prohibited in interstate commerce.[34]

Hearings leading up to the 1938 amendments revealed that the FDA had been "captured" by the pharmaceutical industry, with particular reference to a "revolving door" connecting regulator to industry. Two basic forms of the perceived capture were advanced. The first represents a kind of natural life cycle, which should sound familiar: An agency is formed in response to public pressure, whereupon the pressure recedes. Eventually, the agency strikes a bargain with the regulated industry. The second form of capture is founded in economics. It, too, should strike a familiar note: Either the agency personnel act in their personal self-interest and, in recognition of brief tenure, establish good relations with the industry or policy making is "purchased" by the industry through political alliances with officials of the legislative and executive branches.[35]

The 1938 amendments repealed the 1906 statute. Whatever the impetus, FDA entered into premarket clearance of drugs for safety.[36] Significantly, a drug was automatically approved unless FDA rejected it within sixty days of application.[37] For the first time, incidentally, prescription and over-the-counter drugs were distinguished.[38]

From the thirties on, the government could ban ineffective drugs, but only by proving related claims to be false or misleading, and the burden of proof was with FDA.[39] Congressional attention to the matter of pharmaceutical practices was reawakened in the fifties. Confrontation with the industry was led by hostile, sensational headlines and angry criticism.[40] Among the controversies were indications of rigged research in progress under FDA auspices, questionable practices by physicians paid to test drugs, the presence of "research quacks," and suspicions of physicians willing to "prove" anything they were asked to.[41] But general discontent, however vociferous, is hardly enough. There must always be a focus.

That regulation suitable to the biochemical (and political) realities of pharmaceutical advances in the thirties no longer fit the realities of the sixties can be expressed in one word: thalidomide. This dangerous drug had in fact been kept off our market by successive demands for data while pertinent research was under FDA review. The German originator had also withdrawn thalidomide. Still, it could be used in this country on an experimental basis without any requirement that patients be advised of its status.[42]

Thalidomide clearly revealed gaps in the regulatory testing and tracking of experimental drugs. These gaps occurred at critical junctures, notably in the control of dissemination, in ensuring patient knowledge of its properties, in effecting suitable recalls, and in assuring the maintenance of adequate records by physicians.[43] Dissatisfaction with the regulation of pharmaceuticals therefore coalesced around thalidomide:

> [L]egislation almost certainly would not have ensued were it not for the thalidomide disaster. The [European] epidemic of grossly deformed "seal babies" . . . almost overnight achieved the unanimous passage of [the 1962 amendments]. . . . The fact that thalidomide had not been approved for U.S. marketing was somehow irrelevant, as was the fact that the new requirements . . . , had they been in effect, would not have prevented a thalidomide-type tragedy. The specter of drug toxicity sufficed to carry the day.[44]

The quoted passage conveys a variety of lessons in the context of this book. They will be explored at greater length in this and succeeding chapters.

The hearings leading to the 1962 Kefauver-Harris Drug Amendments were asking questions of considerable pertinence here. They can be summarized in the larger question: Prior to marketing clearance, who is to make decisions regarding pharmaceutical efficacy? The candidates were the recognized experts, the conventional wisdom of practicing physicians, individual practitioners on the basis of patient population, those promoting new drugs, governmental agencies, the public at large, writers of pharmacology texts, and the compilers of drug yearbooks.[45]

Most responsible opinion went against intrusions by the lay public, with an allusion to Krebiozen, a contemporaneous unproved cure for cancer very much in favor among those for whom conventional cancer therapies were ineffective. Practicing physicians were eliminated for lack of time, facilities, skill, and training. By process of elimination, then, experts were suggested as the decision makers of choice, with acknowledgment of their limitations.[46]

But another voice was raised at the same time. Richard Harris observed those who were opposed to the 1962 amendments, comparing them with James Madison's class of persons activated by a common passion adverse to the rights of others or to the "permanent and aggregate interests" of society. Echoing Madison, Harris notes that politicians ignore the broader interests at considerable risk. Methods to do so are indeed available in our time as well as in Madison's but societal interests can be successfully championed, as they were in the amendments of 1962.[47]

In the end, the emergent amendments of 1962 included three major components: provisions addressing the distribution and use of investigational drugs, those calling for evidence that pharmaceuticals are effective as well as safe in their intended usage, and those reforming drug advertising to obtain more balanced statements of their limitations, side effects, and contraindications.[48] With these amendments FDA could ban marketing of new drugs or force withdrawal of existing ones upon finding a lack of evidence of effectiveness, shifting the burden of proof to manufacturers. The standard of proof is substantial evidence, as opposed to either a mere preponderance or an "overwhelming" demand[49] (possibly comparable to the "beyond a reasonable doubt" of criminal law). In addition, automatic approval based solely on the passage of time was abolished. Records of adverse effects were mandated, as were improved quality control and strict FDA inspections of small manufacturers and those manufacturing generics, in the first recognition of generics by FDA.[50] Informed consent on the parts of participants in human clinical trials of pharmaceuticals was also required.[51]

For all the reforms of 1962, problems remain. There are those, for example, who would prefer a free market. While the market has made and will make serious mistakes, FDA's critics perceive little regard on the part of the agency as to its own errors in its efforts to serve society as a whole.[52] Legislative amendment has hardly eliminated disputes regarding standards for determining safety and effectiveness of pharmaceuticals. Such disputes go to the heart of one problem before us: how to apply law to the control of science.[53]

That the regulation of drugs has known its successes is manifested through the abolition of patent medicines, the control over claims and advertising, and the elimination of untested and ineffective medicines.[54] If, however, the mission of pertinent legislation is the elimination of the negative attributes of pharmaceuticals rather than an increase in their benefits, the act is arguably unsuccessful.

Premarket testing has not rendered drugs intrinsically safer, in part because such testing cannot substitute for post-marketing surveillance. Instead of preventing monopolies, this program has yielded a more entrenched position among the largest firms. Moreover, there is a built-in tension among the patent process, lengthy testing demands, and the potential for unjust enrichment. Ironically, the requirement for effective drugs has apparently resulted not in the availability of greater numbers of effective drugs but in an appreciable lag in their marketing in this country.[55] All of these criticisms will be at least implicit in the remainder of this chapter and throughout the next.

In this context and as a background for discussions to follow, an outline of the basic pharmaceutical testing process is instructive. As in any science, knowledge of the literature is the starting point. Out of that knowledge will arise a promising formula or variations on some existing formula. It will be tested for its therapeutic or other beneficial properties and, if the promise holds, for toxicities of its own. If the drug passes both tests, it will proceed to testing in animals, eventually including primates. If results warrant, the drug will then move on to testing in humans, in desperate cases of disease or other morbid conditions or in volunteers. Clinical testing in American patients is the indispensable final step before seeking federal authorization to market, for our purposes, from FDA.[56] This last step reflects the observation that our regulation of pharmaceuticals is more restrictive than that of many European countries,[57] for good and for ill.

ISSUES FOR TODAY

> "Laymen still believe that in the sciences an expert can unerringly tell the true from the false and the good from the bad. Two plus two always equals four, and most people believe the same to be true of any other scientific proposition."[58]

Certainly, today's patient is receiving better value for the money than ever before. That value is in part a product of federal legislation.[59] Some of FDA's caution may be healthful in and of itself. The enthusiasm that follows reports of new pharmaceutical discoveries may need to be tempered. The other side of the coin is that it is possible to ban all drugs and all testing of drugs in the name of safety or in pursuit of the unreachable absolute safety.[60] If there could be a third side to this coin, it would be found with those who would exploit dying patients. FDA goes far in protecting the critically ill from the unscrupulous and unethical.[61]

Such is the tightrope on which FDA is constantly teetering. For example, the pharmaceutical industry's record of profits and of research and development expenditures suggest that these firms and the public interest are simultaneously being served. On the other hand, there is a social cost in the relatively high and inflexible prices of drugs, at least upon introduction.[62]

Some commentators call for examination of collective policy formulation in the pharmaceutical arena. They note the inevitable timorous policies. When a decision is subject to public scrutiny, a regulatory agency will tend toward excessive attention to the worst cases imaginable. With FDA the result has been what is known as the drug lag.[63]

New drug introductions in the United States do seem to be substantially fewer than those elsewhere. From an average of fifty-four introductions per year from 1950 through 1962, the annual rate had dropped to an average of sixteen in the years from 1963 through 1975.[64] A legitimate question is whether the decline is due to regulation or to the attainment of some plateau of knowledge. The answer seems to lie in the numbers: Although such declines are observed on a worldwide basis, the greatest exists in the United States. FDA has argued that this decline reflects drugs with little or no therapeutic advantage.[65]

It can come as no surprise to hear the sound of a reprise from an earlier observer of the bureaucracy. FDA is not immune from administrativitis, the political ailment discussed in the opening chapter. As time passes, the pioneers die out or transfer to the private sector. Those who remain are subjected to constant attack, sometimes insidious. They risk the label of troublemaker, and even the realists grow tired and lapse into the safety of routine. On the outside public interest slackens, while on the inside technical questions bog down the process. New and exciting matters distract from the more routine, and all the while vested interests sustain their campaign of opposition.[66]

What is legitimately to be sought here, as elsewhere, is the conservation and expansion of the benefits before FDA. The attendant danger is found in excesses. Any excess is unacceptable, whether it be in FDA, the pharmaceutical industry, or the medical profession.[67] Even excessive public zeal must be unmasked and viewed with suspicion.

In short, prudence demands that we be aware of what we can and cannot do and what we do and do not know in terms of the medical arts and sciences. For all our skills, we can be ignorant and clumsy. In our quest for truth and healing, we must be wary of self-deception and errors of judgment. Similarly, we demand of the consuming public as individual patients and as a politically potent body the exercise of moderation and prudence. This lesson comes home with a vengeance through the history of the disaster of DES, as recounted in chapter 5, where all acted in good faith and with the best of intentions and within the established medical realities of the time.[68]

The foregoing matters and the difficult issues attending them are manifested constantly. DES is only one manifestation among many with repercussions throughout society and in the forums intended to protect us. In the next chapter, we will examine the issues as they relate to three major contemporary concerns. The common thread will be the influence of regulation on innovation. Incorporated in that thread is the pattern of public participation in the ongoing formulation of policy.

First, FDA's approval process will be explored, with an emphasis on the implications of associated delays in reaching the market with innovative pharmaceuticals. Then we will turn to the difficult matter of generic licensing and its implications for innovation. Finally, we will consider the current status of orphan drugs, those for which there is so limited a market that innovation can become a labor of love without obvious financial reward.

NOTES

1. APPEL, R. and S. FISHER, TO DO NO HARM (1984), at 38.
2. Ibid., at 60–65, *passim*.
3. Ibid.
4. Ibid.
5. T. MAHONEY, THE MERCHANTS OF LIFE: AN ACCOUNT OF THE AMERICAN PHARMACEUTICAL INDUSTRY (1959, reprinted 1972), at 17.
6. Apfel and Fisher, *supra* note 1, at 5.
7. For a popular compendium of the medicinal qualities of plants, *see* MAGIC AND MEDICINE OF PLANTS (The Reader's Digest Association, Inc., 1986).
8. B. BARBER, DRUGS AND SOCIETY (1967), at 9.
9. C. ASBURY, ORPHAN DRUGS: MEDICAL VERSUS MARKET VALUE (1985), at 11.
10. W. DAVIES, THE PHARMACEUTICAL INDUSTRY: A PERSONAL STUDY (1967), at 162, 197. *See also* Mahoney, *supra* note 5, at 1.
11. Gaddam, *A Perspective on Pharmacology*, in DRUGS IN OUR SOCIETY (P. Talalay, ed., 1964), at 22–25.

Double-blind clinical tests are those using patients where neither the physician nor the patient knows whether the subject drug or a placebo (mock drug without therapeutic value) is being administered. In some cases the test drug is compared not to a placebo but to another drug of known therapeutic value. The intent is to eliminate "psychological" responses that might mask the therapeutic properties of the subject drug.

12. Litchfield, *Evaluation of Drugs in Experimental Animals*, in DRUGS IN OUR SOCIETY, *supra* note 10, at 70.

13. W. WARDELL and L. LASAGNA, REGULATION AND DRUG DEVELOPMENT (1975), at 158–159.

14. Apfel and Fisher, *supra* note 1, at 5.

15. Ibid., at 38. There are a variety of materials currently available for both physicians and the curious patient, ranging from inserts in the packaging of prescription and over-the-counter drugs to compendia of pharmaceuticals listing their characteristics, including side effects and other potentials for harm. Of them, THE MERCK INDEX, according to a full-page ad in the February 22, 1991, issue of *Science*, has been published for a hundred years. The centennial eleventh edition includes 10,000 "significant drugs, chemicals and biologicals." Entries new to this edition are "pharmaceuticals with novel mechanisms of action, unique, naturally occurring compounds, and chemicals of environmental concern." This compendium is marketed as a nonprofit service of Merck & Co., Inc.

16. M. STATMAN, COMPETITION IN THE PHARMACEUTICAL INDUSTRY: THE DECLINING PROFITABILITY OF DRUG INNOVATION (1983), at 5–8.

17. Mahoney, *supra* note 5, at 1, 13.

18. See, e.g., Asbury, *supra* note 9, at 36.

19. Ibid., at 48.

20. Mahoney, *supra* note 5, at 16–17.

21. Asbury, *supra* note 8, *passim.*

22. Ibid., at 47.

23. 21 U.S.C.A. 301, *et seq.*

24. Asbury, *supra* note 9, at 34, and Mahoney, *supra* note 4, at 17.

25. Asbury, *supra* note 8, at 63–76, 120 and *passim.*

26. Gaddam, *supra* note 11, at 23.

27. Connor, *The Functions of the Pharmaceutical Industry in Our Society*, in DRUGS IN OUR SOCIETY, *supra* note 11, 117, at 124.

28. *See* M. MINTZ, BY PRESCRIPTION ONLY (rev. ed., THE THERAPEUTIC NIGHTMARE, 1967), at 162–175 *passim.*

29. See, e.g., Kolata, *Heart Institute Is Major Player in Clinical Trials*, 237 SCI. 851, August 21, 1987 News & Comment.

30. H. GRABOWSKI, DRUG REGULATION AND INNOVATION: EMPIRICAL EVIDENCE AND POLICY OPTIONS (1976), at 13–14.

31. Gaddam, *supra* note 11, at 23.

32. Nightingale, *Drug Regulation and Policy Formulation*, 59 MILBANK MEMORIAL FUND Q. (3) 412 (1981), at 442–443.

33. Ibid, at 413.

34. Mahoney, *supra* note 5, at 24.

35. Asbury, *supra* note 9, at 24–25.

36. Nightingale, *supra* note 32, at 413.

37. Ibid., at 414. *See also* Grabowski, *supra* note 30, at 14.

38. Asbury, *supra* note 9, at 13.

39. Cutler, *Practical Aspects of Drug Legislation*, in DRUGS IN OUR SOCIETY, *supra* note 11, 149, at 150.

40. Wardell and Lasagna, *supra* note 13, at 1.

41. R. HARRIS, THE REAL VOICE (1964), at 218–219.

42. Asbury, *supra* note 9, at 25.

43. Ibid., at 26.
44. Wardell and Lasagna, *supra* note 13, at 1.
45. Goodman, *The Problem of Drug Efficacy: An Exercise in Dissection*, in DRUGS IN OUR SOCIETY, *supra* note 11, 49, at 65–67.
46. Ibid., at 65.
47. Harris, *supra* note 41, at 140.
48. Goodrich, *The Responsibilities and Problems of Government*, in DRUGS IN OUR SOCIETY, *supra* note 11, at 141. A useful synopsis of the Federal Food, Drug, and Cosmetic Act is presented in M. WOROBEC and G. ORDWAY, TOXIC SUBSTANCES CONTROL GUIDE (1989), at 64–76. This guide includes a brief history of this legislation, notably including amendments in the eighties establishing an orphan drug program, to be discussed in chapter 6.
49. Cutler, *supra* note 39, at 151.
50. Asbury, *supra* note 9, at 28. *See also* Nightingale, *supra* note 31, at 416 and *passim* for more on the provisions of the 1962 amendments.
51. Flannery, *Should It Be Easier or Harder to Use Unapproved Drugs and Devices?*, 16 HASTINGS CENTER REP. (1) 17 (1986). Ellen J. Flannery is an associate at the Washington, D.C., firm of Covington and Burling.
52. Grabowski, *supra* note 30, at 11.
53. Cutler, *supra* note 39, at 150.
54. Wardell and Lasagna, *supra* note 13, at 127.
55. Ibid., at 128.
56. Mahoney, *supra* note 5, at 21. Certain authorizations are the province of the National Institutes of Health or the American Medical Association's Council of Drugs.
57. Cutler, *supra* note 39. Cutler provides a detailed review of the process of reviewing new drugs and points out the associated hazards facing interested parties, not the least of whom are those potentially served—or disserved!
58. Ibid.
59. Wardell and Lasagna, *supra* note 13, at 127.
60. See, e.g., Cutler, *supra* note 39, at 157–158.
61. Flannery, *supra* note 51, at 23.
62. Markham, *Economic Incentives and Progress in the Drug Industry*, in DRUGS IN OUR SOCIETY, *supra* note 11, at 176.
63. Wardell and Lasagna, *supra* note 13, at 140. *But see also* Nightingale, *supra* note 32, at 439–440.
64. Grabowski, *supra* note 30, at 18.
65. Ibid., at 19–20.
66. Mintz, *supra* note 28, at 416–420.
67. Ibid., at 384–385 *ff.*
68. Apfel and Fisher, *supra* note 1, at 129 and *passim*. DES is also reported in Asbury, *supra* note 9.

CHAPTER 5

Regulation, Voices of the Public, and Pharmaceutical Innovation: Implications of Delays in FDA Approval of New Drugs

When FDA control of the entry of new drugs into the marketplace is reviewed in the historical context of pharmaceutical advances, we confront a disturbing possibility. Could the pharmaceuticals we take for granted even reach the stage of clinical trials under the aegis of contemporary review?[1]

THE FDA APPROVAL PROCESS

> "[I]t is premature to adopt a rigidly legalistic approach to matters of drug efficacy and safety. . . . The answer is to ensure that the . . . regulations do not overstep the current capabilities of science."[2]

We have noted in passing that amendments to the Federal Food, Drug, and Cosmetic Act[3] have emerged from an ongoing furor surrounding the practices of the Food and Drug Administration (FDA). The furor has arisen from increasing sophistication regarding pharmaceuticals among the medical professions and from increasing public scrutiny of medicinals. In this context it is instructive to recall the long history of the botanical origins of therapeutic substances and to keep in mind that almost half of the pharmaceuticals we presently take for granted are still derived from those botanical sources.[4] That even such drugs are a mixed blessing has always been apparent.

A prime example is deadly nightshade, the source of belladonna. Many a child has succumbed to the sweet berries of this plant, but derivatives of natural belladonna are used to dilate the eyes for medical examination and are muscle relaxants and sedatives. These derivatives, too, are deadly poisons. Extremely toxic even in small doses, they can cause coma and death. Another familiar example is foxglove, used therapeutically to regulate heartbeat despite the risk of heart failure due to its toxicity. Both properties have been known since the late eighteenth century.[5] Despite their hazards, these substances are used to good effect and without controversy, perhaps because they are natural as opposed to manufactured. Only in this century have we learned to synthesize pharmaceuticals, although chemical pharmacology can be traced to the Renaissance Swiss chemist, Paracelsus, who recognized the importance of chemical principles in the preparation of drugs and of dosage in therapy.[6]

Lacking any theory behind drug effect or toxicity in humans,[7] we have turned to individualized regulation for assurances of both efficacy and safety in the drugs we take. But for those assurances, FDA and the medical professions must turn to scientific testing, itself subject to regulation. The balance to be struck in such regulation is fine indeed. Because relevant science and technology continue to advance but will probably never be entirely responsive to the queries raised, current law tends toward broad statements and interpretations more generous yet. At the same time that law and science are being forced into a working team, we seek constantly to improve the science in aid of more suitable decisions.[8]

To this end we make certain assumptions and turn to a central authority, FDA, for a determination of safety and effectiveness before a new drug can enter the marketplace. The assumption is that current definitions of safety and effectiveness are adequate despite a lack of adequate criteria and some demand for a statistical rather than a clinical approach. We further assume FDA to be capable of clear distinctions between appropriate and inappropriate pharmaceuticals and that animal screening and pre-market clinical tests will clearly guide those distinctions. In application of these principles, we effectively pit society against the individual patient.[9]

The process FDA pursues is in two parts: Investigational New Drug (IND) and New Drug Application (NDA), both subject to regulations.[10] The IND process commences with a Notice of Claim for Investigational Exemption of New Drug. There are three principal tests: (1) determination of safety in healthy volunteers, (2) review of safety and efficacy in a small number of patients, and (3) confirmation of safety and efficacy while establishing optimum dosage in large numbers of patients. A fourth testing phase is in post-marketing clinical trials.[11]

For the most part the actual review process is relatively informal between FDA and the pharmaceutical applicant, with industry tending to avoid more formal procedures. Industry recognizes public sensitivity to the slightest official suggestion of hazard. Thus, informal suggestions of disapproval will lead to acquiescence on the part of an applicant. While this practice is entirely appropriate to the early stages of the process, more formal approaches later on could define issues, specify grounds for FDA decisions, and provide a basis for review that is fully meaningful.[12]

This process is extraordinarily time-consuming. Once a drug has been isolated, preclinical trials can take from one to two years. Chronic testing in generations of animals then proceeds while human trials commence. These phases of testing can last some five years, for a total approaching ten years.[13] In 1967 associated costs were approaching half a million dollars for initial animal studies leading to the decision to test in humans. Costs were approaching a million dollars for initial clinical testing leading to submission of the NDA and exceeding a million dollars from the point of NDA to FDA approval,[14] which might ultimately be withheld.

To avoid a mistaken assumption that the informality suggested above represents laxity on the part of FDA, it is essential to acknowledge the inherent power of this agency. FDA personnel, in interacting with sponsors of new drugs, may exercise qualitative and quantitative judgments having potentially profound effects on drug development. Furthermore, agency personnel would not be human if they could avoid implications of retaliation regarding present or future filings if displeased with an applicant's behavior.[15] Indeed, the agency is also armed with a "terrifying armory of legal weaponry" to encourage compliance without ever resorting to enforcement through injunction, seizure of products, or criminal prosecution, any of which can be quite isolated from the fact of a subject drug's safety or efficacy.[16]

One might reasonably assert that FDA should be not a policeman but more an agent assisting in bringing safe and effective drugs to the marketplace as soon as possible.[17] Indeed, there may be merit to the assertion that drugs are being unnecessarily withheld from the public by the very agency charged with protection of the public health vis-à-vis pharmaceuticals. Some might never reach clinical trials. Botanical derivatives such as digitalis and quinine are cited as examples of drugs in current use that might never be cleared today.[18]

Two questions are thereby raised: Is the public being served, and what does the public have to say about the situation? If a single set of answers exists, they require yet another query: Where do the voices of the public enter the fray associated with the approval process and attendant delays in moving a promising pharmaceutical from initial isolation to prescription? Significantly, neither the questions nor their answers are necessarily consistent with one another. Yet, that the public really does drive FDA's approval process cannot be doubted. The public drive is a general one in the sense that federal legislation and regulation are responsive to demands of the electorate. The drive can also be splintered for and against a selected administrative matter—or detoured by a specific therapeutic agent that has caught the public imagination.

THE VOICES OF THE PUBLIC

> "Despite our increasingly sophisticated knowledge of long-term side effects of drugs and the new focus of efficacy, are we any better protected

> from the same tendency toward consensus of public, academic, and commercial interests that existed with DES?"[19]

It has been said that we really do not know whether we should be grateful for the regulatory price being paid to avoid hasty introductions of innovative pharmaceuticals or distressed by the needless suffering attributable to the preclusion of therapeutic agents from the market. Moreover, as this quandary persists, no one is charged with finding an answer or with comparing alternatives.[20]

Perhaps as a result, the drug industry is estranged from many segments of the public. At best, the public is confused or poorly informed. At worst, there is outright belligerence. The estrangement itself has been attributed to lurid headlines.[21] The pressure to move a new drug into the marketplace can be intense, especially in situations deemed desperate. Among these are the plight of parents who have lost their babies during pregnancies and of women desperate to conceive. Such pressure can lead clinical investigators with any sensitivity to retreat from the equally real need for proper trials of potentially effective drugs.[22] This tension is often stated in dramatic terms: How many could have been saved by a drug kept off the market for however long? The presumption is that a more rapid approval process would save lives. But the counterpoint is, how many lives could have been saved by regulatory delay, when a dangerous drug manages to slip through the intense scrutiny.

There are numerous cases in point. Any "scientific" answer would lie with properly framed queries subject to rigorous statistical process,[23] but perceptions are arguably more important in terms of policy formulation. Public voices will be raised where interest is provoked and are loudest where the greatest interest can be found. A great provoker is fear. Today, the fears of individuals for their personal health are most likely to be centered on reproductive capacity and on three diseases of the body and one of the mind: cancer, heart disease, AIDS, and Alzheimer's disease. Because perceptions are paramount here, those are the situations on which we will focus.

DES, A Harbinger of Things to Come

> "[The lesson to be learned] is not who to blame or what technical procedure[s]... avoid the errors of our predecessors, but that we are subject to the same stresses and temptations as the actors in the DES tragedy."[24]

A devastating and representative case is that of DES, diethylstilbestrol. This introductory case in point reveals ongoing relationships among public perceptions, as conveyed by the various voices of the public, and the responses of the medical community and FDA itself. It is important to realize that, if there are no heroes here, neither are there any villains.[25] All were pursuing the same mission, and one still before us: to speed up investigations of promising drugs while prohibiting from the marketplace unsafe or ineffective drugs, whether the source be the pharmaceutical industry, government, or academia.[26]

The management of the drug DES by all concerned presents a number of object lessons, perhaps still to be completely absorbed. The drug is in the first instance representative of the timeless conflict at the intersection where the medical profession and the public, together seeking out bold new advances, meet more conservative elements of society who hold the conviction that medical arts are most effective when nature is allowed to take its course.[27]

DES met an accurate perception of need. The time was the 1930s. All the interested sectors were in search of an inexpensive, easily produced synthetic estrogen to support endangered pregnancies. But DES proved to be more toxic as well as more effective than its natural counterpart. Carcinogenicity was established in the forties, and the drug was banned from veterinary uses. Nevertheless, it was approved for medical uses, essentially at the same time. Those few physicians who opposed approval were seen as impediments to progress.[28]

Many of the questions raised with DES are heard today. Ironically, they often occur in the context of cancer therapies, a reversal of the DES situation. Going beyond the quality of pharmaceutical research, these questions arise from conflicting interests represented by consumers, physicians, the pharmaceutical industry, and administrative agencies. In the case of DES, the drug was not initially approved for any use during pregnancy but was later declared safe without any regulatory constraint on its usage.[29]

By the fifties, when double-blind studies were conducted, DES was actually found either to be without effect or to worsen problems. Unfortunately, however, the consuming public had been "educated" by lay articles and only the most courageous practitioner could refuse it to a patient. As a result, its use was widespread, even in normal women. In short, DES availability exemplifies the power of anecdotal reporting over more formal research. Proper research would have prevented any use of DES during pregnancy as both ineffective and unsafe.

The first, otherwise rare cancer in DES daughters appeared in 1970, with administrative alerts following in the next year. By then DES was routinely used in "quality" care bestowed upon predominantly middle-class patients, notably including the wives of physicians. DES was even incorporated in vitamins intended for pregnant women. As an additional irony, the resultant lawsuits brought by the daughters inflicted with cancer following prenatal exposure to DES taken by their mothers have been not against the prescribing doctors but against the pharmaceutical manufacturers.[30]

DES was subject to FDA review, at least to the extent that the process existed prior to the 1962 amendments. None of the participants, however, was fully aware of the degree to which respective positions were governed by special interests, prevailing medical fashion, and even passions. The politicolegal system failed the ultimate victims, because that system was defeated by the enthusiasms of the medical establishment and of the public itself.[31] The caveats are that, in the matter of DES, all acted in good faith with the best of intentions. Behavior was within the established medical norms of the day. In the end, it seems such consequences are unavoidable.[32]

Cancer, A Quandary for Today

> "The public posture . . . is that, if the substance is 'harmless' . . . *and* if an individual . . . is terminally ill, the patient has a right to receive medication as long as it is relatively safe."[33]

Today the passions of the public have resulted in other accommodations. For cancer, FDA has established a category for innovative drugs developed by the National Cancer Institute. This category, comparable to existing "Compassionate IND" or "Treatment IND," makes drugs available before review is completed. The affected patient's informed consent is required before the drug can be administered. In fact, one FDA committee is responsible for development of a systematic approach to such compassionate mitigations of approval processes.[34] The problem is that the public, through terminal patients almost certainly beyond the help available through accepted therapies, demands access to unproven therapies, those outside of or apparently languishing in FDA's approval process. Among these are yesterday's laetrile[35] and today's Neupogen[36] and IAT, "immuno-augmentative therapy."[37]

Of these three therapies, Neupogen is the one that has proceeded through regular FDA channels. It is nontraditional only in that it is a product of recombinant DNA biotechnology. This treatment to reduce life-threatening infections associated with cancer proved so effective in clinical trials that ethics demanded termination of double-blind studies to treat more patients, another indication of external pressures on the FDA approval process. Here, circumstances demanded an acceleration of the approval process, probably legitimate in terms of safety and effectiveness as well as ethical considerations. In this case the pressures are consistent with FDA's overall mandate, assuming that side-effects remain more inconvenience than hazard.[38]

Laetrile was studied not under FDA auspices, but through the National Cancer Institute (NCI). The substance was derived from a toxic component of peach pits. Rejected by the medical community as a hoax in the 1970s, laetrile became a notorious sociomedical issue of national scope. Never approved by FDA, laetrile was legalized in more than half the states and used by 75,000 patients, rivaling accepted chemical treatments, before dropping out of the headlines and presumably out of the lives of the desperately ill.[39]

At the time NCI was initiating its study of IAT, there was no scientific evidence of its safety or effectiveness. Only one practitioner was promoting its use. The American Medical Association did not endorse it. IAT promotion had not assumed the proportions of the laetrile uproar. Why, then, was the therapy even under consideration by this agency? One answer is that the Congress, the surrogate of the public, was demanding the study, essentially at the instigation of a single representative. The circumstances here are problematic but hardly uncommon. The sole practitioner of the therapy may be charlatan or prophet. We have no way, short of following through with studies at public expense, of distinguishing the two within the politicolegal system.[40]

A Broader Perspective

> "[D]espite all the publicity . . . ,the parallel tracking system has not yet been formally instituted."[41]

Comparable problems have arisen with regard to innovative treatment of blood clots leading to heart attack,[42] novel therapies for the treatment of AIDS,[43] and breakthroughs in the treatment of Alzheimer's disease.[44]

Such situations mark the desperation of terminal patients and reveal FDA's difficult task of providing meaningful protection without impeding true medical progress. That task can only be more difficult in face of public voices, especially where more impassioned than informed. Within the aegis of current approval processes, sectors of the public (or their surrogates, elected or self-appointed) can become participants in the course of clinical trials, in the actual approval process, and in seeking out alternatives to the existing bureaucratic process, as was the case with laetrile and IAT.

While reform may be the mission, a defeat of the purpose behind regulation is also a reasonably predictable result. Demonstrations of the safety and effectiveness of innovative drugs may never be clearly established. In effect, regulation of innovative pharmaceuticals regresses to the situation preceding the 1938 amendments. Public pressure, which initiated politicolegal reforms of 1938 and 1962, can itself obviate those very reforms. When the inevitable failures result, and a new DES or thalidomide emerges, there are the substantial damages to be collected in court from manufacturers and the allegations of incompetency and worse leveled against regulators. According to one principal investigator, the question comes down to whether sound scientific method will prevail or will give way to activism.[45]

From the perspective of Alzheimer's comes another voice, loud and clear. Underlying the essentially political process responding to research into treatment of this disease is a grim reality, the incalculable suffering of patients and their families. This suffering is accompanied by expenditures of tens of billions of dollars annually, eaten up by patient care extending for as long as the fifteen years the condition can persist.[46] Only a sampling from these cases in point need be described to explore the nature of this multifaceted conundrum beseiging policy formulation.

Starting with the political process itself, we find FDA castigated no matter what decision it reaches. When a committee of FDA declined to approve Genentech's Activase, a tissue plasminogen activator known popularly as "TPA," a generalized outrage followed.[47] There was speculation that FDA's analysts did not understand the scientific issues.[48] The quality of Genentech's data was questioned by some, while others insisted the agency had focused on inappropriate questions.[49] Still others cited an internal turf battle at FDA or simple bureaucratic inefficiency.[50]

From the media came the question, "Is U.S. medical research being conducted for the benefit of patients or for the benefit of the research system and the approvers?"[51] The scientific community was agitating along the same lines:

> [A]ny appreciable delay in approving a drug widely considered to be a drug of choice is not simply a bureaucratic minuet—it is a matter of life or death to many patients. It is therefore incumbent on an agency to ask truly scientific questions and not simply go through pro forma experiments that were appropriate [some twenty years in the past].[52]

Finally, the role of individual personalities cannot be dismissed. Individuals are an inevitable part of the system,[53] but they add another layer of complexity to controversies already complicated in technology and policy.

Turning from the political process to the scientific, we encounter another issue, equally difficult: the ethics of using placebos. Arguably essential to the validation of pharmaceutical efficacy as well as safety, placebos represent a heartbreaking impasse. Nowhere is it more intense than in cases of children suffering from AIDS. That those receiving a placebo in place of a promising experimental drug will die is as close to fact as any scientific probability can come. A closely related problem is that the test drug itself can be dangerous to the very same victims of this scientific process.[54] No one can do "the right thing" in such a situation. FDA and all the participants are wrong whether the drug proves to be dangerous or both safe and effective. Either way children are going to die with a consequent public outrage.

Although it seems unlikely that double-blind tests with placebos can be eliminated, on more than one occasion recognition of the ethical dilemma has resulted in an interruption of testing to take patients off a placebo when the test drug is particularly promising. What follows, however, is a demand that the test drug then be made widely available before testing is complete.[55]

If the scientific and bureaucratic communities do not move quickly enough for those suffering from devastating disease, one sector of the public, the patients themselves, will take matters into their own hands. This sector will in effect make its own scientific and political decision out of a potent mix of desperation and frustration.[56] At that point there is little the politicolegal system can do to improve the situation for either the patient or the public at large. Recent history shows that the resultant disasters will be remedied in the judicial system[57] and in renewed calls for legislative reform.

Reaction at the Intersection

> "If exceptions are made for AIDS, the same arguments would hold for cancer and other currently incurable diseases afflicting millions of people."[58]

Defying the hazards, there is today a partial bureaucratic response to the quandary. It is the "fast track" or "treatment IND," also known as the "parallel track." This fast track has been instituted out of disagreements among elected officials, bureaucrats, and informed scientists, often rendered highly emotional by public demands, also at odds. The process suggests a system confounded to the detriment of scientific and technological progress. The situation is an embarassment to the politicolegal system, and it is all proceeding at the expense of the most vulnerable participants of all, individual patients of the present and the future.[59]

The fast track itself is not without controversy. It is a manifestation of a very real tension between the basic need for reliable clinical information and the desire to afford hope to dying patients. This tension was explored by a variety of public and private sectors in a conference held in 1990. The shift toward providing hope carries its own irony: Those patients will be assuming "far more of the risk of unproved drugs."[60] In addition to the disruption of all elements of pharmaceutical testing represented by this pending approach, aversion to risk is a current politicolegal reality, to which FDA is presently firmly attached:

> [A]ll systems that allow expanded access to drugs before they are finally approved have this in common: They increase the risk of toxicity or unexpected side effects in the patient population. Until now, the FDA philosophy was to minimize risk to all patients absolutely, but AIDS patients have taken the lead in saying that they want the option of taking some increased risk if the payoff is access to a potentially effective agent.[61]

A seemingly unavoidable conclusion is that policy on access to pharmaceutical innovations is to be driven by "gross civil disobedience."[62] An editor of the *Wall Street Journal*, Daniel Henninger, is of the opinion that FDA, in response to pressure from AIDS activists, will move quickly to their demands and then close the door on researchers and advocates for other diseases. He offers in explanation:

> Historically, the FDA has been obsessed with risk. This is so because over the past 75 years occasional, highly publicized accidents involving drugs have given congressmen a pretext to announce, with cameras rolling, that the agency is too sloppy with risk. Whether the general public ever feared risk as much as the politicians claimed during these controversies is a good question.[63]

The media have been very much involved, but not necessarily with any consistency (or factual grounding). A transfer of power to the patients can overwhelm science. Moreover, there is what may be an artificial wall being erected between today's patient and the interests of society itself. Henninger asserts that, out of this melee of interests, the dominant view in the coming decade is that benefit deserves equal status with risk. More significantly, he asks who will be deciding the issues. Henninger turns away from both Congress and the FDA to another agency, Health and Human Services (HHS), and the presidency,[64] both legitimate surrogates of the public, but limited in perspective and ability or willingness to integrate the issues in the context of the array of underlying facts.

In any event, we have now come full circle. Voices of the public demanded and got first safety and then effectiveness. Now those voices are demanding access, with little regard for the niceties of scientific or legal proof of either. There is no particular pattern to successes or failures in politicolegal responses to the demands.

The overall situation arising from delays in approval of innovative pharmaceuticals can be stated succinctly. Basic disagreements inevitably occur among informed scientists, bureaucrats, and elected officials. The effect of such disagreements is compounded by public demands. The established approval system is thereby confounded to the detriment of scientific and technological progress, the embarrassment of the politicolegal system, and—the greatest irony of all—at the personal expense of the most vulnerable participants, those suffering from the most devastating inflictions.

PERCEPTIONS AND IMPLICATIONS

> "There is profound ignorance of [the process of drug development] in academic, professional, and government circles, and in the media. Individuals in all these communities should be encouraged to grasp the nature of the drug development process at all levels and the principles on which it is based."[65]

Heart attacks kill half a million Americans a year. At the close of 1987 AIDS had killed some twenty-one thousand Americans. A drug to combat death from heart attack could not get FDA approval. One to alleviate the symptoms of AIDS moves on FDA's new fast track.[66] Which is the right course? We may never know. For different reasons both may prove right or both wrong. The most discouraging point is that crucial decisions are being made in response to the loudest voice, with diminished regard for the broader view, integrating the full array of considerations—scientific, political, economic, ethical, social, and psychological—with individual needs.

The specter of another laetrile or, worse, another thalidomide was not exorcised in 1938 or in 1962. Whether and how we can correct the situation remain problematic questions. Politicolegal history would seem to hold out little hope. The drive of the public is inconsistent. This matter of access to innovative pharmaceuticals may have the largest following, but it is not the only pertinent issue of pharmaceutical innovation. The one overriding consistency is in an absence rather than a presence. In 1975 there was no comprehensive, balanced analysis of what should be required of the interested sectors or how all the competing requirements were to be met.[67] The same can be said as we enter the nineties.

In the preceding case, a reasonable answer seems to lie in existing FDA policy, which tends to be lost in the public uproar. In fact, it is asserted that the fast-track approach is not extraordinary, although it is so colored through publicity. If the agency is authorized to continue to allow the process when a patient does not respond to more traditional treatment, all interests could be served. After all, it is FDA's goal to make such drugs available on the basis of need, once safety and efficacy are indicated. In the absence of these data, patients may be subjected to worthless investigational drugs while foregoing any opportunity to take another that might offer better hope for control of their disease. FDA itself can direct physicians and patients to the latter drugs. But the agency can educate the public only by codifying its practice through promulgation of regulations.[68] Clearly, a distinction, as one participant observed, must be made between headache remedies and a drug for one threatened with imminent death.[69] Is this too fine a distinction to ask of the politicolegal system and the marketplace, the partnership whose raison d'être is service to the American public?

In the next chapter, the companion matters of generic and orphan drugs will be scrutinized. The contexts, not necessarily compatible, continue to be pharmaceutical innovation, its control, and the public interest.

NOTES

1. Litchfield, *Evaluation of Drugs in Experimental Animals*, in DRUGS IN OUR SOCIETY (P. Talalay, ed., 1964), 69, at 75.
2. W. WARDELL and L. LASAGNA, REGULATION AND DRUG DEVELOPMENT (1975), at 136.
3. 21 U.S.C.A. 301 *et seq.*

4. MAGIC AND MEDICINE OF PLANTS (Reader's Digest Association, Inc. 1986), at 5. This eminently readable volume provides a brief history of medicine and medicinals at pages 49 through 73.

5. Ibid., at 96, 188.

6. Ibid., at 51–58, *passim.*

7. BARBER, DRUGS AND SOCIETY (1967), at 9.

8. Wardell and Lasagna, *supra* note 2, at 136.

9. Ibid., at 136, 140 *passim.*

10. Nightingale, *Drug Regulation and Policy Formulation*, 59 MILBANK MEMORIAL FUND Q. (3) 412 (1981), *passim.* The regulations are codified at 21 CFR Parts 312 (IND) and 310 (NDA).

11. C. ASBURY, ORPHAN DRUGS: MEDICAL VERSUS MARKET VALUE (1985), at 29–34.

12. Cutler, *Practical Aspects of Drug Legislation*, in DRUGS IN OUR SOCIETY, *supra* note 1, 149, at 152–153.

13. Asbury, *supra* note 11, at 31–34.

14. Barber, *supra* note 7, at 122.

15. Wardell and Lasagna, *supra* note 2, at 19.

16. Cutler, *supra* note 12, at 153. Furthermore, FDA can make an allegation stick and can invoke sanctions exceeding the magnitude of the alleged violation. The chances of an applicant's prevailing in administrative or judicial hearing are considered slim and rendered something of a nullity, given the industrial fear of publicity. In short, according to Lloyd M. Cutler, FDA can impose its will by the lifted eyebrow. Ibid.

17. *See* H. GRABOWSKI, DRUG REGULATION AND INNOVATION: EMPIRICAL EVIDENCE AND POLICY OPTIONS (1976), at 7.

18. Litchfield, *supra* note 1, at 75.

19. R. APFEL and S. FISHER, TO DO NO HARM (1984), at 19. The authors are psychiatrists who were counselling victims of DES.

20. Ramo, *Regulation of Technological Activities: A New Approach*, in SCIENCE AND LAW: AN ESSENTIAL ALLIANCE (W. Thomas, ed., 1983), 35, at 39–40.

21. Folkers, *Drug Research in Industry*, in DRUGS IN OUR SOCIETY, *supra* note 1, at 131.

22. Apfel and Fisher, *supra* note 19, at 23.

23. The media are not constrained by statistics in the same sense scientists and the bureaucracy are. For example, consider this excerpt: "Weeks matter. Every time a week passes, another 9,000 Americans die of heart attack. It has been 23 weeks since [disapproval of a promising new drug], which works out to some 200,000 heart-attack deaths. TPA wouldn't have saved them all, but it would have saved some." *Days and Nights at the FDA*, WALL ST. J., Nov. 11, 1987 editorial, at 36.

More-objective reports are available. One estimates that twelve hundred lives might have been saved during the five years required for U.S. approval of a drug already available in Great Britain. Grabowski, *supra* note 18, at 16. On the other hand, an equally difficult problem is removal from the market of drugs found to be worthless. Usually, such drugs are expensive, and they can deflect patients from potentially life-saving therapies. Flannery, *Should It Be Easier or Harder to Use Unapproved Drugs and Devices?*, 16 HASTINGS CENTER REP. (1) 17. *See also* Reffenbach, *Unproven Remedies Tempt the Ailing*, WALL ST. J., May 27, 1988, at 19.

24. Apfel and Fisher, *supra* note 19, at 130.
25. DAYS AND NIGHTS, ibid, at 8–9.
26. Asbury, *supra* note 11, at 80.
27. Apfel and Fisher, *supra* note 22, at 5.
28. Ibid., at 11–18.
29. Ibid., at 18–19.
30. Apfel and Fisher, *supra* note 22, at 20–25, 87–88.
31. Ibid., at 127–128.
32. Ibid., at 130.
33. Nightingale, *supra* note 10, at 437.
34. Asbury, *supra* note 11.
35. *See*, Nightingale, *supra* note 10, at 437. *See also* Holden, *NCI May Evaluate "Alternative Therapy,"* 241 SCI. 1285 (1988).
36. *See* Rundle, *FDA Approves Amgen Cancer Drug, But Rivals Loom*, WALL ST. J., Feb. 22, 1991, at B1.
37. *See* Holden, *supra* note 35.
38. Rundle, *supra* note 36.
39. Holden, *supra* note 35.
40. Ibid.
41. Skerrett, *Parallel Track: Where Should It Intersect Science?*, 250 SCI. 1505, Dec. 14, 1990 News & Comment.
42. Aspects of the public perspective on the chronicle of clot-dissolving drugs can be followed in *Science* or in the *Wall Street Journal. See* e.g., Sun, *FDA Puts New Heart Drug on Hold*, 237 SCI. 16, Jul 3, 1987; *The Decision against a Heart Drug* (excerpts from the FDA committee meeting of May 29) and *The Flat Earth Committee* (related editorial), WALL ST. J., Jul. 13, 1987, at 20; Koshland, *TPA and PDQ*, 237 SCI. 341, Jul. 24, 1987 editorial; *TPA: Tip of the Iceberg*, WALL ST. J., Dec. 2, 1987 editorial, at 28; Sun, *The Coming Competition among Clot-Busting Drugs*, 240 SCI. 1267, Jun. 3, 1988; Hudson, *Genentech's Heart Drug TPA Appears Only to Equal Its Rivals, Report Says*, WALL ST. J., Sep. 2, 1988, at 34; Marx, *Which Clot-Dissolving Drug Is Best?*, 242 SCI. 1505, Dec. 16, 1988.
43. From a comparable perspective the chronicle of AIDS research can be followed. *See* e.g., *Today AZT, Tomorrow TPA*, WALL ST. J., Mar. 23, 1987 editorial, at 18; Chase, *For Some AIDS Patients, Bootleg Drugs Are One Way to Preserve Some Hope*, WALL ST. J., Oct. 5, 1987, at 20; Ricklefs, *Medical Newsletters on AIDS Therapies Crop Up across U.S.* WALL ST. J., Oct. 4, 1988, at B4; Bacon, *Plan to Speed Availability of Drugs for AIDS Backed by U.S. Officials*, WALL ST. J., Jul. 21, 1989, at B2; Marshall, *Quick Release of AIDS Drugs*, 245 SCI. 345, Jul. 28, 1989; Marx, *New AIDS Drug Passes First Clinical Test*, 245 SCI. 353, Jul. 28, 1989; *Wider Use of AIDS Drugs Advocated*, 245 SCI. 811, Aug. 25, 1989; Palca, *AIDS Drug Trials Enter New Age*, 246 SCI. 19 Oct. 6, 1989; Cherfas, *AZT Still on Trial*, 246 SCI. 882, Nov. 17, 1989; Palca, *New AIDS Drugs Take Careful Aim*, 246 SCI. 1559, Dec. 22, 1989; Chase, *FDA's Peck Steps in at Crucial Time for AIDS Research*, WALL ST. J., Feb. 11, 1991, at B1.
44. Alzheimer's research is also being reported from public perspectives. *See*, e.g., Marx, *FDA Queries Alzheimer's Trial Results*, 239 SCI. 969, Feb. 26, 1988; *Policing Alzheimer's*, WALL ST., J., Mar. 3, 1988 editorial, at 16; Sattin, *THA Takes on Alzheimer's and FDA*, WALL ST. J., Apr. 11, 1988, at 26; *Wichita Falls Calls the FDA*,

WALL ST. J., Apr. 29, 1988; Marx, *NGF and Alzheimer's: Hopes and Fears*, 247 SCI. 408, Jan. 26, 1990.

45. Marx, *Drug Availability Is an Issue for Cancer Patients, Too*, 245 SCI. 346, Jul. 28, 1989. The investigator is Charles Moertel of the Mayo Clinic. A relatively obvious additional question is whether a *Science* or a *Wall Street Journal* or the two in combination should stand for the voice of either the public or reason. On the other hand, is there *any* institution outside government that can do so?

46. Sattin, *supra* note 44.

47. *See* note 42, *supra*.

48. Sun (1987), *supra* note 42, at 16.

49. Ibid., at 17.

50. Ibid., at 18.

51. "Flat Earth," *supra* note 42.

52. Koshland, *supra* note 42.

53. *See*, e.g., "Policing Alzheimer's," *supra* note 44, and Chase (1991), *supra* note 43. Significantly, in 1988 Dr. Frances Kelsey, who once held up approval of thalidomide, was working to defeat the sole researcher-proponent of a questionable drug for the alleviation of Alzheimer's symptoms. One cannot help but wonder just whose side the public may be taking—and whether in hindsight Dr. Kelsey will be justified or vilified. There would appear to be no middle ground.

54. *See*, e.g., Chase (1987), *supra* note 43.

55. *See*, e.g., Cherfas, *supra* note 43.

56. *See*, e.g., Chase (1987), *supra* note 40, and Booth, *An Underground Drug for AIDS*, 241 SCI. 1279, Sep. 9, 1988.

57. Such is the case with Accutane, the "thalidomide" of the eighties. *See*, e.g., Lammer, *Accutane and Public-Health Risk*, WALL ST. J., Jun. 16, 1988 letter to editor, at 23; Blodgett, *Acne Control vs. Birth Defects*, ABA J 17, Jul. 1, 1988; Barrett, *Lawsuits over Accutane Start to Mount*, WALL ST. J., Aug. 3, 1988, at 21. Apparently, FDA had decided to allow the manufacturer to "police" distribution of Accutane, in an effort to avoid a lengthy court battle. The agency cited prior efforts to regulate previously approved drugs, only to be thwarted by the judicial system. Waldholz, *FDA Unlikely to Demand Restrictions on Accutane*, WALL ST. J., May 10, 1988, at 41.

See also, Ruling That Drug Didn't Cause Birth Defects Upheld, WALL ST. J., Sep. 1, 1988, at 2 (Benedectin). Such lawsuits themselves call into question the multidisciplinary issues and their management by the judicial arm of the politicolegal system. *See, Drowning Drug Research*, WALL ST. J., Dec. 1, 1988 editorial, at A22.

58. Marshall, *Quick Release of AIDS Drugs*, 245 SCI. 345, Jul. 28, 1989 News & Comment.

59. *See*, e.g., Bacon, Marshall, and Rickleffs, *supra* note 43.

60. Skerrett, *supra* note 41.

61. Ibid., at 1506.

62. Henninger, *Will the FDA Revert to Type?*, WALL ST. J., Dec. 12, 1990, at A16. Daniel Henninger is deputy editor of the JOURNAL's editorial page.

63. Ibid.

64. Ibid.

65. Wardell and Lasagna, *supra* note 2, at 155.

66. *First Do No Harm*, WALL ST. J., Dec. 21, 1987 editorial, at 20.

67. Wardell and Lasagna, *supra* note 2, at 158.

68. Flannery, *supra* note 23, at 20–21. Ms. Flannery was writing as an associate at the D.C. law firm of Covington and Burling. This issue was widely reported in 1987 and 1988. *See*, e.g., Chase, *FDA Rule Changes May Rush New Drugs to Very Sick Patients*, WALL ST. J., Oct. 10, 1987, at 1; *FDA Rule Changes on Breakthrough Drugs*, WALL ST. J. Oct. 29, 1987 letters to editor, at 33; Booth, *FDA Looks to Speed Up Drug Approval Process*, 241 SCI., Sep. 16, 1988, at 1426; *FDA to Shorten Testing Stage for Some Drugs*, WALL ST. J., Oct. 20, 1988, at B4; Waldez, *Drug Firms Hope FDA Broadens Plan to Speed Approval of Some Medicines*, WALL ST. J., Oct. 21, 1988, at B3.

69. Booth, Ibid.

CHAPTER 6

Regulation, Voices of the Public, and Pharmaceutical Innovation: The Quandary Goes On

THE EFFECT OF GENERIC LICENSING ON INNOVATION

> "Indeed delays [in FDA approvals] are a way of life, in part because FDA reviewers have much to lose if they allow an inferior drug on the market and little to gain for speedy work."[1]

In 1975 there were over eight hundred manufacturers of prescription pharmaceuticals. Of these, the top thirty held 85 percent of the market. The smaller firms tended to sell mostly under generic names without investing in significant research and development.[2] Before this discussion proceeds, some legally recognized distinctions are in order.

Pharmaceuticals are known by a variety of names. A chemical name describes a substance's basic structure. The same chemical substance is likely to be better known by a trademark, a trade name, or a brand or generic name. Trade names and trademarks are legally recognized identifications of suppliers and drugs. Brand names are less precisely defined legally, but are generally the names best known to consumers. All three are "proprietary," that is, subject to legal protection for specific purposes. Generic names are those considered "nonproprietary" common names. Ordinarily, less formal (and less daunting) than a chemical name, a generic identification is owned by no one. Anyone can use it.[3]

By custom, physicians are generally inclined to prescribe by brand or trademark, the philosophy being therapeutic reliability with recognition of slight differences among "identical" drugs. Even when the question of cost is raised, the traditional inclination is more often to move to less expensive brands rather than to generics.[4] Arguably, to say the least, it is necessary to assure full equivalency in prescribing a generic in preference to a brand. Unfortunately, equivalency, another issue for science and technology, is not readily ascertained for regulatory purposes.[5]

Here, it is helpful to pause again to consider another politicolegal reality. A patent, that legal device by which unique pharmaceutical formulations are protected by the federal government, has a lifetime of only seventeen years. The brevity of this legalized monopoly represents a very real limitation for pharmaceutical innovations. Once that patent issues, time is running. It is essential to secure FDA approval well within the lifespan of the monopoly. The duration of the patent following approval determines the effectiveness of the monopoly, which commences upon entry into the market and concludes with patent expiration. Upon expiration, corresponding generics may begin to compete. FDA's approval process has effectively been reducing the interval of patent protection since 1950.[6]

Trademark can effectively extend the period of corresponding market dominance. But other federal and state law will determine the availability of generic counterparts and thus alter the structure of the marketplace. That body of law, while external to substantive issue of pharmaceutical policy, represents still another trade-off between low prices and incentives for innovation.[7] The initial societal cost-benefit ratio associated with the research and development of a drug cannot be expected to carry beyond the juncture where generics or other substitutes are available.[8]

Generics and "branded" generics, the latter bearing a firm's name and reputation, have been available since the fifties. In fact, investigations leading up to the 1962 amendments included generics, which were rarely prescribed at the time.[9] But generics were not widely acceptable until the generic pharmaceutical industry was favored with 1985's Waxman-Hatch Act, expediting their approval.[10]

A related question arises with regard to who is in control of the pharmaceuticals being prescribed by physicians and purchased by patients. The question also relates to guarantees of quality. No doubt, some combination of internal (marketplace) and external (politicolegal) control is necessary. The issue, of relative degree, relates to the manner in which the interested publics are participating in the overall process.[11] Significantly, while neither a brand name nor a generic is indicative of quality,[12] there is a public perception to that effect. Obviously, additional public perceptions will also figure into each situation.

It is informative to use the rivalry between the manufacturers of name-brand pharmaceuticals and those of their generic counterparts to illustrate the ways in which participants in policy formulation can take advantage of elements of the politicolegal system or engage in outright abuse. A scandal reported in the eighties serves to reveal how one indispensable participant, in this case the bureaucracy, can be an accomplice to those seeking advantage in the marketplace. In most cases, it is simply a matter of a party's availing itself of favorable intricacies of the politicolegal system. In others, however, there are misuses, at least in terms of either patient benefits or some societal "good." From use, it is a short reach to abuse. At least in the cases going no further than mild misuse, there can be every appearance that sound science and legal due process are being observed. (On the other hand, experience shows that sound practices of information exchange or negotiation may be perceived as somehow suspect.)

As of 1985, with the enactment of the Waxman-Hatch amendments, approval of generics was based on bioequivalency with their brand-name counterparts. Simultaneously, numerous patents expired; and concern about the extraordinary costs of medical care prompted Medicaid, state law, and private insurers to prefer generics unless a doctor prescribed to the contrary. As a result of the juncture of these events, generics rose from less than a tenth of the market to more than a third, while the number of prescriptions remained the same.[13]

Despite the market expansion, there was anecdotal evidence suggesting differences between brand-name pharmaceuticals and their generic counterparts. Some doctors persisted in demanding brand names, especially when dosages were particularly sensitive.[14] There was, however, no scandal until 1987.[15]

During 1987 a manufacturer of a generic drug was anticipating the first FDA clearance of that drug. In a business representing nearly nine billion dollars a year, first clearance meant almost two billion dollars. First clearance, however, unexpectedly went to another, and the disappointed applicant was prompted to investigate through detectives. Evidence of bribery of key FDA personnel was uncovered and taken to congressional investigators and federal prosecutors.[16] By 1989, possibly falsified bioequivalency data were disclosed, and FDA was moving to revoke tainted approvals.[17]

For the most part this ongoing scandal was out of the public view simply because it fostered no headlines. Even so, such battles affect every physician, numerous popular medicines, countless patients, and medical insurers. In the eighties, the fight turned ferocious and often damaging to the public. Having underwritten scientific studies damning the safety of generic drugs, the big firms used them to frighten state bureaucrats and lawmakers into restricting generic drug use. At the same time, many of the generics makers gave as good as they got.[18] Any inference that this debate is no more than petty squabbling is quickly dissipated with recognition of the dollars at stake, especially when patents in some major brand names are due to expire in the interval of 1991 through 1995.

In addition to these major components of a scandal, there was a suggestion that relevant educational programs were being subverted through funding programs among those involved. In 1989 FDA was reportedly prompted to write to one of the participants to assert that it had "employed falsehoods, misrepresentations, and innuendo on a continuing basis and in the absence of any valid scientific data, in an effort to convince the medical and pharmaceutical communities that [the recipient's product] is superior"[19] to all alternatives. Whatever the outcome and its ramifications, this "drug war" was expected to escalate during the nineties.[20]

Where can the public voices be found in this situation and to what effect? The voices can be found within FDA itself and in communication with that agency. They are heard in the Congress and in the legislative bodies of the several states. They are reflected in the marketplace, in the news and entertainment media, and, when all else fails, in the judicial system. In short, they echo and re-echo throughout our society. As the sound becomes increasingly distant from its sources it becomes distorted, but as policy evolves the resultant perceptions sustain new voices having greater import than their better-informed counterparts.

There are a variety of ways in which these voices are heard or distorted and may influence other sectors. The list begins in FDA itself, initially as its potent legal tools are implemented. Though every tool utilized by the agency may be wholly within applicable law, each lends itself to manipulation resulting in everything from minor inconvenience to members of the regulated community to outright wrongdoing that may rise to a level vulnerable to criminal charges.[21]

At the outset, FDA has a power going beyond initial approvals of innovative or generic pharmaceuticals. FDA can also revoke or suspend existing approvals. All the related authority represents a powerful force indeed. These internal processes, well within the power of administrative authorities, can lead to very different perceptions on the part of participants and outside observers. For example, are agency demands for more data and changes of mind legitimate, or is something more sinister at work? Perceptions vary with the observer, as was the case during the scandal in the eighties.

Perceptions are instructive, in terms not only of FDA itself but also of any program where policy formulation is comprised of a complex mix of politics, law, science, technology, and societal demands. Stir in the mass media, and the recipe is truly threatened by multitudes of cooks who may or may not be in full cognizance of what it is they are concocting.

Among the perceptions are a lack of consistent internal rules, guidelines, or policies for implementing agency authority. This perception is accompanied by those associated with the inevitable changes in approval processes and their management. Some believe existing rules are altered or ignored on whim. That belief is only reinforced by a perceived intrusion of individual philosophies mutually exclusive and potentially extraneous to FDA's mission. Perceptions of hidden private agendas abound.

Others may assert that review assignments are selectively related to reviewer work-practices, which speed or impede the review process depending on the identity of drug or applicant. Moreover, reassignments of personnel can coincide with corresponding alterations in approaches to a process, especially suspect when imminent approval is thereby delayed.

Also of concern are such tactics as initiating an investigation without providing any rationale to affected markets, which can to potentially inappropriate reactions among interested sectors. For example, files can be lost with resultant delays in the ongoing regulatory process.

All of these problems, real or imagined, are exacerbated when reactions to assertions of incompetence or deliberate manipulation provoke an ineffective response. Further exacerbation follows from even an implied threat of retaliation for informal or formal complaints raised within or outside the agency, particularly among those who fear discrimination following any appearance before congressional committees overseeing FDA and its operations.

Moving from executives to legislative forums, we find a variety of voices exerting a variety of influences in Congress and in the legislative bodies of the several states. In the abstract these voices are entirely legitimate. They take two forms: the provision of information in response to pending legislation or ongoing investigations and frank lobbying. A potent tool for both is the letter campaign. But this process, too, along with its component parts, can be misused or subjected to outright abuse. The result is pressure that is at least unethical, if not illicit or even criminal. How are the policymaker and associated experts to distinguish among information, disinformation, propaganda, and outright lies? If the so-called insiders are hard-pressed, what of the affected and more remote sectors of society? Can there be any wonder that cynicism mars our view of government and marketplace?

Moving into the marketplace and closer to the general public, we find pharmacists removing one drug from their shelves in favor of another in response to an ongoing FDA investigation. Suspicion that something may be wrong is enough to prompt such action. Quite naturally, competitors will seek a marketplace advantage when one of their number is under FDA scrutiny. Competition, as always, can go too far, at the expense of some array of sectors.

When the processes of the regulating community and the marketplace are hopelessly at odds, the third branch responsible for governmental policy formulation enters the fray. In the case of the FDA scandal of the eighties, evidence was provided to federal prosecutors, who ultimately brought criminal charges against agency personnel. Civil lawsuits were also initiated against FDA personnel and members of the regulated community alleged to have participated in illicit manipulations of agency processes. Antitrust and racketeering allegations rounded out the appeals to the judiciary.

Separately, in the competition between pharmaceutical innovators and their generic competitors, the benefits of approval for either can be impeded through the patent process. Delayed approval can defeat a patent. A patent infringement suit can delay the entry of a competing drug.

Thus, the outright wrongdoing illustrated here can be as extreme as acceptance or solicitation of bribes by FDA personnel and outright cheating on the part of applicants or their competitors. Sorting out the realities of these complex matters will be no easy task, especially when the process is inextricably bound up with arcane science and technology.

POLICY FORMULATION REGARDING ORPHAN DRUGS

> "Although drugs for rare diseases and unpatentable drugs may have been unattractive to industry prior to the 1962 amendments, these and other orphan-drug categories began to be publicly acknowledged by industry as particularly unwise investments after the amendments were passed."[22]

What Are Orphan Drugs?

> "Many drugs turned up in the course of research can never become profit makers."[23]

Orphan drugs are drugs effective against diseases suffered by so few people that the proprietor cannot make a profit from their sale and or even recover the costs of research, development, and marketing. To make such drugs available to those who need them, the manufacturer must sell them at a reduced profit or as a no-profit service.[24] Since 1962 only two introductions of orphan drugs per year are reported, most of them expected to serve less than a hundred thousand persons in this country. Because of the research and development behind them, some of these drugs are extremely expensive. The initial question, then, is whether orphan drugs should be developed without regard to their cost. If at least some are not to be developed, who will decide where to draw the line on cost? Who will pay whatever costs are incurred in the process?[25]

There are numerous examples of orphan drugs, including one specifically effective against botulism and others effective against such diverse hazards as phenylketonuria and snakebites.[26] Others are effective against conditions associated primarily with the Third World, desperately needed but themselves in need of monetary subsidies.[27]

Actually, orphan drugs come in a variety of categories in addition to those characterized by a small market. Other categories include drugs that, for one reason or another, represent a limited market in combination with major liability, such as childhood vaccines and drugs taken by women during their childbearing years. Unpatentable drugs are also considered orphans, as are those that treat "unpopular" conditions such as addiction or intended for undeveloped nations.[28]

All these are orphans because of the substantial risks, financial or legal, associated with their development and distribution. Many enjoy little or no government support, on the proposition that they should not be underwritten by public funds. At the same time, academic support is limited because of prohibitive liability insurance.[29] Also known as "public service drugs," these orphans may be made available by their manufacturers in a spirit of public service or as a matter of public prestige. But, obviously, only financially strong manufacturers can take such a step, and even the strongest can be inhibited either by perceptions of excessive regulation or threats of liability.[30] Additional disincentives are the high costs of raw materials, requisite individualization of administration for each patient, whose needs will be changing, and the requirement to educate physicians.[31]

Significantly, in the context of this book, FDA anticipates new biotechnology techniques to expand the arsenal of orphan drugs, especially for diseases related to genetics, metabolism, and the endocrine system in addition to neurological disorders and cancers.[32] Here, the scientific, political, and public voices heard at FDA, and beyond, will be in chorus with their counterparts at the Environmental Protection Agency. Dissonance is hardly unexpected.

There can be little doubt that the legal framework surrounding pharmaceutical innovation is itself an impediment to the availability of orphan drugs, especially in the aftermath of the 1962 amendments. The combination of low profits, extensive regulation, and widely publicized litigation (with or without an ultimate finding of liability) can be devastating.[33] For example, some drugs in common use have not been tested for efficacy in rare conditions for fear of adverse effects associated with those rare conditions. Such effects could unnecessarily affect regulation of the more common use.[34]

In short, as little as a decade ago the sources of support for orphan drugs were simply inadequate. Patients were not reimbursed for drugs that were investigational. Both insurers and the government required FDA approval of drugs before costs were reimbursed. Pharmaceutical interest was restrained. Relevant programs of the National Institutes of Health were incomplete and unlikely to expand. Although there were legislative activities, consumer advocates were unresponsive to this issue.[35]

To reiterate, the questions before policymakers are whether these orphan drugs should be developed notwithstanding associated costs, who should bear the burden of development, what should determine that a drug is not to be developed. Any answers can have general application to other major issues affecting interactions between public and private sectors.[36]

Public Needs and Public Demands Regarding Orphan Drugs

> "[I]deas in good currency are those that are powerful enough to stimulate public policy."[37]

In response to the need for orphan drugs, the Pharmaceutical Manufacturers Association established a Commission on Drugs for Rare Diseases. The Department of Health and Human Services (HHS) set up an Orphan Products Board, and the FDA established its Office of Orphan Product Development. Separately, volunteer health agencies instituted NORD, the National Organization for Rare Diseases.[38] National interest in the problem was not piqued, however, until an article appeared in the *Los Angeles Times* and an episode of the television series *Quincy* highlighted the issue.

Between them, these highly public items placed orphan drugs in "good currency," a position that has its obvious benefits but is not without its detriments:

> Such ideas start at the periphery and often, by the time they arrive in the mainstream, no longer accurately reflect the existing state of affairs. [One conclusion is] that a principal criterion for effective learning is the social system's ability to reduce this lag time so that ideas in good currency do reflect current problems.
> ... [Before a House hearing actor Jack Klugman] reenacted from the witness table his congressional appearance from [*Quincy*] to argue on behalf of legislation for orphan drugs. ... [The actor] took industry and government to task in full public view before a crowded press section and television cameras.
> ... Before this dramatic stimulant to public awareness of orphan drug problems, however, much had transpired to lay a foundation for action.[39]

Perhaps the most effective long-term response to the matter of orphan drugs is that of NORD. This organization lists eight objectives. Notable among them are the establishment of a clearinghouse for distribution of information on rare disorders and on orphan drugs; fostering of communication among volunteer agencies, government, industry, researchers, academia, and individuals; education of the general public as well as of the medical community; and focusing of attention on the needs of patients suffering from rare disorders.[40]

Even though FDA asserted as nearly as 1977 that mechanisms were available to expedite review of orphan drugs, agency personnel were unable to provide guidelines in point, because the processes were informal and varied with circumstances. Clearly, more formality was in order, despite the limitations associated with governmental formalities.[41]

Response through Policy Formulation

> "Within a year, the plight of orphan drugs, presented in a balanced yet dramatic show, was captured from an insider's dialogue ... to a nationally recognized tragedy."[42]

Whatever the impetus or its level of legitimacy in terms of societal benefits, the matter of orphan drugs has been taken up by Congress. The atmosphere in the early eighties has been described by one observer as one of listlessness in the health industry under the "lackluster" leadership of physicians and scientists and subject to a government called both "cavalier" and "callous" with regard to pharmaceutical entrepreneurship as manifested through innovation.[43]

The first legislation to address orphan drugs was introduced in 1980, following a decade of administrative studies. Another bill was introduced at the end of 1981, and the Orphan Drug Act was enacted in January of 1983.[44] But it can come as no surprise that interest and controversy persisted beyond enactment.

Reauthorization of the law was anticipated in 1988, at which time funds available for research grants providing lifesaving drugs to sufferers of rare diseases were to be increased. New requirements regarding specification of orphan drugs were to be imposed and the tenure of the National Commission on Orphan Diseases extended.[45]

The success stories linked to the Orphan Drug Act are heartwarming, but this law is not without its critics. When reauthorization loomed in 1990, an unlikely coalition of AIDS activists, biotechnology firms, and politicians were pursuing changes to preclude certain members of the industry from earning large sums at the expense of other interested persons. Some interest was also attributed to taxpayers at large, in that some $100 million in taxes were going into just one orphan drug. The manufacturer in question was reportedly making that same amount on the product.[46] As the controversy raged, each critic was moved by individual reasoning, but it all came down to financial considerations. It does seem reasonable that a drug that brings in millions of dollars is no orphan, at least as contemplated by the movement culminating in the statute. The bone of contention is the temporary (and limited) exclusivity vested in one manufacturer of an orphan. One representative of public interests, NORD, was wary of any change, noting that, even with protection, sponsors of orphan drugs are not easily found.[47] As a spokesperson for NORD put it, "[W]e are absolutely opposed to any form of shared exclusivity because it would cut the heart out of the Orphan Drug Act. . . . We are unwilling to sacrifice the fate of 20 million Americans with orphan diseases because AIDS patients want less expensive drugs."[48]

And so, those who formulate policy on behalf of the public are faced not only with differing positions on science and technology but also with public voices in direct conflict with one another. Where there is no obvious way to accommodate opposing interests, the question can become how to make alterations in policy without subverting it in the process.[49] Reauthorization sponsor Congressman Henry Waxman had an answer, at least in the abstract:

> The act should continue to apply with full force to the orphan drugs that have only one sponsor. It would continue to stimulate research on drugs for rare diseases. But

where a drug's profit potential is so great that it has many parents competing to invest in its future, it is no orphan and should not have protection against price competition.[50]

NORD agreed with the congressman, as did the research institution (and orphan drug sponsor) Memorial Sloan-Kettering Cancer Center. A manufacturer of one orphan drug, however, asserted that exclusivity was critical to its ability to attract needed investment for breakthrough therapy.[51]

Later in the same year, an AIDS patient asserted that, although the act had extended his life, proposed amendments would defeat the provisions in point. According to this voice of the public, the pending amendments would signal pharmaceutical firms that Congress would not stand by its word, at least where controversy arises. This patient concurred with the representative of the manufacturers in asserting that the ability to retain profits is crucial to innovation for orphan diseases and that retroactive lifting of orphan status would be counterproductive.[52]

In fact, the compromise amendments ultimately reached were subjected to a pocket veto. It seems that the President agreed with this voice on behalf of one kind of patient and of one sector within the manufacturing community.[53] Contrary voices spoke out elsewhere with a more complete review of the circumstances at hand.[54] Of them, one observed that the controversy culminating in the pocket veto represents a typical paradox in the successes of our system to which the government must be alert. This voice, however, offers no resolution of the paradox to be effected by the government or any other formulator of policy.[55]

CONSOLIDATION AND SYNTHESIS

"The vital functions of research and discovery . . . can be harmed by well-meaning individuals with responsibilities outside the laboratory."[56]

It would seem that FDA and other formulators of policy, for all their respective efforts to do the right thing, are perpetually wrong, in the eyes of one sector or another. The most generous reading of the situation at FDA suggests that the agency is working in full compliance with the letter and spirit of the laws for which it is responsible, with a sublime disregard for the realities of the affected publics, most notably the pharmaceutical industry and its market. A more cynical reading is that FDA personnel, out of purely personal missions, are deliberately manipulating those laws to the benefit of some applicants and the detriment of others. The most cynical will see behavior that is unethical and even criminal. Few are likely to affirm that any national mission is being realized to effect the best for the individual in keeping with societal needs at large.

Unfortunately, it can be all but impossible to discern where the lines should be drawn, whatever the forum. Distinctions can be extremely subtle and clear only to the very insiders who are being viewed with suspicion. Wherever the lines may fall, it is the patient and, ultimately, society as a whole who are thereby disenfranchised in this extremely personal and significant area of public policy and law.

Issues of Public Participation

> "Most of all I hold it the responsibility of the thinking men and women in this country to inform themselves intelligently and clearly on pending legislation and make themselves articulate."[57]

In our society the public legitimately drives both the marketplace and governmental policy, especially social regulation. In the case of control of pharmaceutical innovation, certain perceptions come into play. Many cling to a perception that "natural" products are somehow more to be desired than those of human manufacture. In the case of medicinals, both classes represent the same benefits and hazards in kind and degree.

In the arena of pharmaceutical availability, public pressures create a tightly strung tension linking the need for testing with that for rapid approval of promising innovations. Yet another tension can be exerted by pressure to keep the marketplace free of drugs that are dangerous or ineffective. These tensions are only aggravated when certain elements of pressure are based in ignorance or misinformation. No one is entirely free of responsibility here, but media attention can carry a limited perspective far enough to skew priorities and to effect perceived reforms inappropriate to the actual problem.

Resultant disillusionments endanger the nation's precarious balancing of governmental and marketplace control. On occasion balance is lost, and efforts will be ill-conceived or misdirected altogether. This distraction from legitimate missions can occur in the private and public sectors and crosses all the lines we so carefully draw to distinguish among professions and to isolate policy from improper influences.

Examples of these difficult realities abound. Here it is enough to reiterate only a few. In the course of debate preceding the 1938 amendments, one woman, a consumer, was demanding a law with real teeth in it. She sought the protection of public health and safety without any intrusion into "honest healthy spots" recognizing that such intrusion could defeat the regulatory mission, she wanted protection against abuse rather than interference with proper usage.[58] On the same day, another representative of women's interests was asserting that no governmental proposal seeking to protect public health could be considered either too expensive or overly drastic.[59] An earlier witness, also identifying with "the public," had characterized the pending legislation as diabolical. He believed it would rob "man of ownership" and launched a diatribe against the medical community at large.[60]

Obviously, a wide-ranging sample of public voices were being heard, some providing detailed criticism in combination with precisely drafted proposals for revision. There were even those who recognized that public demands were often coupled with profound public ignorance. Taken as a whole, the statements and the observations they prompted reveal more than the immediate problem and associated perceptions of the time. They also suggest the issues that would arise in the future and have been looking through the eyes of those who have followed.[61]

By the early sixties, when amendments addressing pharmaceutical efficacy were pending, sophistication had expanded the horizons for participation. Over the span of years there was, however, no reduction in the conflicts among interested voices. Conflicting perceptions of the needs had not lessened, and by then the press had become a presence to be reckoned with or to be turned to an advantage.

Senator Estes Kefauver was sensitive to the deficiencies of the politicolegal system and reportedly did his personal reputation little good through his methods, but he was skillful in invoking publicity in advancing his legislative agenda.[62] According to Richard Harris, a contemporaneous commentator, Kefauver had a genius for the role of publicity. He knew the reporters present had little time or space to describe the detail of a full-blown congressional hearing. The items would instead focus on certain aspects of the process. For example, according to Harris, the senator would time certain events to occur close to reporters' deadlines. Moreover, "he often intervened to make a point that no one else had thought of... and repeatedly moved in at a critical moment to restate a point in a way to make it perfectly clear to the press." He also knew the value of body language and studied modulation of voice.[63] As Harris noted, "Kefauver's approach was open to fair criticism. Congressional hearings have invariably been described by those who conduct them . . . as nothing more or less than objective attempts to get at the truth, but the reality doesn't appear to be that simple. They are rarely unbiased."[64]

Harris quotes a member of the press to the effect that congressional investigations are seldom truly investigations. Instead they are quite deliberate strategies to " 'move from a preconceived idea to a predetermined conclusion.' "[65] Kefauver himself noted the need to stir people up about an issue, because of the relative weakness of the consumer as against the strength of corresponding special interests,[66] although he admitted the possibility of "extravagant abuse."[67]

Nearly twenty years later, official reports derided the "glacial pace" of drug approval, exacerbated by the combination of circumstances reported here. At the outset, FDA's mandated regulatory approach was a conservative one. Conservatism was only abetted by a relationship between FDA and the pharmaceutical industry described most generously as adversarial. Moreover, both regulators and the marketplace were being subjected to the intense scrutiny of Congress and the public.[68] As has been suggested, at least the effect of the 1962 amendments has been the subject of more than one indictment.[69] Some thirty years after passage, a small pharmaceutical firm asserted that the statutory provisions at issue could not have been drafted by anyone who knew the necessity to balance the risk of procedural (not substantive) failure against the benefits of innovation.[70]

A major problem, yet to be resolved, is that in large part the public is being "educated" by lay articles. It takes much courage on the part of a responsible party, such as a physician, to refuse demands arising from that education.[71] The question then becomes whether an overriding influence in policy formulation can be attributed to nonrepresentative voices. It would appear the answer must be in the affirmative.

What Are the Remedies?

> What difference does it make if there is not a public demand, if there is a public need?"[72]

It has been observed that medical research is a delicate matter. The public interest is best served when the control of such research is negotiated in a spirit of mutual confidence and respect.[73] Although this observation has considerable merit, there is another participant, who must be prepared for the roles being assigned. That is the public itself. With technical virtuosity and sophistication at least theoretically present, a kind of generalized human excellence must be added. This call for excellence is nothing new. Aristotle called it practical wisdom, or prudence. In more recent times, it was called judgment. In the twentieth century it is identified as foresight. In sum, it is the human capacity to make independent, autonomous judgments about urgent matters not yet subject to general rules, policy formulation for the present and future.[74]

Where Are We Going?

> "The public needs to be kept informed by the scientific community in order to make intelligent decisions on public policy and it has the right to know how public funds are being spent. The problem is to minimize as far as possible the disadvantages of publicity, to see that the public is indeed informed, not misinformed."[75]

Given the array of voices and the propensity of the politicolegal system to respond to the loudest voices rather than to the integrated voices of the informed, what is wanting is an omniscient overseer. This overseer would be charged with the compilation and coordination of all the positions. The mission would be proposal of the single approach that best serves both the public at large and the sectors more immediately affected.

This statement of mission and strategy is an ideal. No realist will expect or demand an ideal of the real world, but surely we can come closer to realization of the ideals embodied in the control of pharmaceutical innovation than is possible through manipulations effected through wildly fluctuating, uncoordinated emotional reactions.

There are no clear mechanisms by which we can sort out fact from perception or educate the public accordingly, so that the ideals of foresight in partnership with judgment and practical wisdom might be realized. We seem incapable of dissipating the web of false impressions and are bound to some plateau in our response to national trauma impelling pharmaceutical innovation, where distortions and revenge are salient motivators. How are we to promote that lofty goal characterized by mature reintegration in the service of realistic and effective public policy?

NOTES

1. Stricharchuk, *Frustrated Drug Firm Charges Corruption by Officials of FDA*, WALL ST. J., Jun. 9, 1989, at 1.
2. M. STATMAN, COMPETITION IN THE PHARMACEUTICAL INDUSTRY: THE DECLINING PROFITABILITY OF DRUG INNOVATION (1983), at 3.
3. BRANDS, GENERICS, PRICES AND QUALITY: THE PRESCRIBING DEBATE AFTER A DECADE (Pharmaceutical Manufacturers Association (PMA) (1971), at 2. PMA represents manufacturers of brand name pharmaceuticals. *See* Winslow, *Generic Drug Scandal Creates Opening*, WALL ST. J., Sep. 6, 1989, at B1. The counterpart for manufacturers of generics is the Generic Pharmaceutical Industry Association. *See* Waldholz and Stricharchuk, *Fight between Generic and Major Drug Firms Heats Up As Stakes Rise*, WALL ST. J. Oct. 4, 1989, at 1.
4. PMA, ibid., at 10.
5. Ibid., at 32–42 and *passim*.
6. Statman, *supra* note 63, at 62–63.
7. Ibid., at 65.
8. Markham, *Economic Incentives and Progress in the Drug Industry*, in DRUGS IN OUR SOCIETY (P. Talalay, ed., 1964), 163, at 176.
9. ASBURY, ORPHAN DRUGS: MEDICAL VERSUS MARKET VALUE (1985), at 24.
10. Winslow, *supra* note 3. The Waxman-Hatch Act (Drug Price Competition and Patent Term Restoration Act of 1984) in pertinent part amends 21 U.S.C.A. 355 and 505 to allow for abbreviated New Drug Applications for generics.
11. B. BARBER, DRUGS AND SOCIETY (1967), at 128–129.
12. PMA, *supra* note 3, at vi.
13. Winslow, *supra* note 3.
14. Ibid.
15. Stricharchuk, *supra* note 1.
16. Ibid.
17. Winslow, *supra* note 3.

18. Waldholz and Stricharchuk, *supra* note 3.
19. Ibid., at A20.
20. Ibid.
21. The following listings in the text are extracted and blended with some inferences from the articles identified in notes 1 and 3, *supra*.
22. Asbury, *supra* note 9, at 38. There appears to be a consensus to the effect that the amendments themselves are in large part responsible for reduced development of orphan drugs. Ibid., at 2 and 21.
23. T. MAHONEY, THE MERCHANTS OF LIFE: AN ACCOUNT OF THE AMERICAN PHARMACEUTICAL INDUSTRY (1959, reprinted 1972), at 28.
24. Ibid.
25. Asbury, *supra* note 9, at 5, 141, 198.
26. Mahoney, *supra* note 23, at 28.
27. Waldholz, *Merck to Donate Drug for "River Blindness,"* WALL ST. J., Oct. 22, 1987, at 42.
28. Asbury, *supra* note 9, *passim*.
29. Karch, *Why Orphan Drugs?* in ORPHAN DRUGS (F. Karch, ed., 1975), 4, at 4–5.
30. DRUGS IN OUR SOCIETY, *supra* note 8, at 291. The specter of liability may even exceed the limited market as a disincentive. Asbury, *supra* note 9, at 87–109, 120 and *passim*.
31. Goldberg and Zarolinski, *Development of Dopamine: An Example of Academic and Industrial Collaboration, in* ORPHAN DRUGS, *supra* note 29, at 117–137 *passim*.
32. Asbury, *supra* note 9, at 83.
33. Ibid., at 2.
34. Ibid., at 81.
35. Karch, *supra* note 29, *passim*.
36. Asbury, *supra* note 9, at 5.
37. Ibid., at 111
38. Ibid., at 4. NORD, with headquarters in New Fairfield, Connecticut, publishes a newsletter (*Orphan Disease Update*) and provides a networking program for sufferers of rare diseases and others interested in this matter.
39. Ibid., at 111–112.
40. Undated NORD brochure received by the author in 1987.
41. Asbury, *supra* note 9, at 179.
42. Ibid., at 111.
43. DeFelice, "The Carnitine Story," *in* Karch, *supra* note 29, at 34. Stephen DeFelice was associated with Bio/Basics International and the Foundation for Innovative Medicine.
44. Asbury, *supra* note 9, at 33, 118–129, 157. PL 97–414 is codified at 21 U.S.C.A. 360.
45. 6 ORPHAN DISEASE UPDATE (3) 3 (1988).
46. Gibbons, *Billion-Dollar Orphans: Prescription for Trouble*, 248 SCI., May 11, 1990, at 678.
47. Ibid., at 678–679.
48. Ibid., at 679.
49. Ibid.

50. Waxman, *Orphan Drug Act Needs Discipline*, WALL ST. J., May 23, 1990 Letters to the Editor, at A23.

51. Meyers, Quirk, and Binder, *Orphan Drug Act Needs Discipline*, ibid. *See also* 8 ORPHAN DISEASE UPDATE (3) 1 (1990). The NORD article reported the statute's success in terms of more than three hundred drugs designated as orphan and more than forty of these approved by FDA. More approvals were pending at the time of publication. This article provides a succinct overview of FDA's approval process.

52. Foster, *Orphan Drug Act Isn't Broken—Don't Fix It*, WALL ST. J., Oct. 18, 1990, at A16.

53. Gibbons, *Orphan Drug Compromise Bush-Whacked*, 250 SCI., Nov. 16, 1990 News & Comment, at 905.

54. *President Vetoes Orphan Drug Amendments* 8 ORPHAN DISEASE UPDATE (9) 1 (1990); Thoene, *Curing the Orphan Drug Act*, 251 SCI., Mar. 8, 1991 Policy Forum, at 1158.

55. Thoene, ibid., at 1159.

56. Folkers, *Drug Research in Industry*, in DRUGS IN OUR SOCIETY, *supra* note 8, 131, at 132.

57. Statement of a woman consumer, in A LEGISLATIVE HISTORY OF THE FEDERAL FOOD, DRUG, AND COSMETIC ACT AND ITS AMENDMENTS, Volume 1, Dec. 8, 1933, 430–435, at 435.

58. Ibid.

59. Statement of a representative of a women's association, LEGISLATIVE HISTORY, *supra* note 57, 448–450, at 449.

60. Statement of a representative of the National Liberties Association, LEGISLATIVE HISTORY, *supra* note 57, 224–225 *passim*.

61. LEGISLATIVE HISTORY, *supra* note 57, *passim*.

62. W. DAVIES, THE PHARMACEUTICAL INDUSTRY: A PERSONAL STUDY (1967), at 60.

63. R. HARRIS, THE REAL VOICE (1964), at 58.

64. Ibid., at 70–71.

65. Ibid., at 71.

66. Ibid.

67. Ibid., at 72.

68. Asbury, *supra* note 9, at 24–25. This relationship is strained and of questionable benefit to the public, because FDA sees itself as a police force dealing with criminals. The apparent goal is to keep drugs off the market, rather than to foster those that are safe and effective. H. GRABOWSKI, DRUG REGULATION AND INNOVATION: EMPIRICAL EVIDENCE AND POLICY OPTIONS (1976), at 6.

69. Grabowski, ibid., *passim*.

70. Homsy, *FDA: Life-or-Death Decisions*, WALL ST. J., Jan. 25, 1988 letter to editor, at 31. Charles A. Homsy identified himself as a consumer of medical services as well as the president of Vitek, Inc.

71. R. APFEL and S. FISHER, TO DO NO HARM (1984), at 23.

72. Observation of Senator Copeland, LEGISLATIVE HISTORY, supra note 57, at 372.

73. Davies, *supra* note 62, at 73.

74. Apfel and Fisher, at 129.

75. DRUGS IN OUR SOCIETY, *supra* note 8, at 297.

CHAPTER 7

Regulation, Voices of the Public, and Products of Recombinant DNA: Science and Technologies

BIOTECHNOLOGY

> [S]imply to understand some aspect of nature . . . can be neither good nor bad. . . . [I]t is not the knowledge itself that contains moral value, but the effect of that knowledge on the individual . . . [in relation to] experiences, values, and beliefs."[1]

The term "biotechnology" represents a broad concept and can be unnecessarily intimidating to the uninitiated. Defining biotechnology, then, can offer some reassurance. The definition can take one of two directions, and both go far toward eliminating any mysticism associated with the concept. First, biotechnology covers any situation in which a science interpreting biological systems is applied to a technical or industrial process. Second, biotechnologies can be seen as being enabling technologies for other technologies.[2]

One familiar form of biotechnology provides clarification: For centuries the human species has been selecting for desirable traits in plants and animals. We call this process selective breeding. It continues today as we select for improvements in domestic species of animals and plants.[3] Additional clarification of the term can follow from a brief recitation of its history as a context for the phenomenal advances of the last few years.

History of Biotechnologies

> "Humanity has been using genetic engineering since the first grandmother saved the biggest bean seed for next year's crop and a more docile wolf was kept as a camp animal."[4]

As technologies evolve, biotechnologies have come to include any technique using living organisms or components of living systems to improve plants or animals. Through such techniques, microorganisms have long been developed for specific uses. But in more recent years the term *biotechnology* has expanded into the manufacture or modification of a broader assemblage of products.[5] As has already been described in some detail, progression from science to technology is logical, seemingly inevitable, and undeniably useful. Beyond meeting needs directly, technology development also represents substantial financial investment, plays a major role in national economics, and invokes political values.[6] Nevertheless, the same history reveals the impossibility of even imagining what the future will really hold.[7] This is surely the case with the current techniques for genetic selection at the molecular level rather than through selective breeding at the level of whole organisms.

The ancient techniques of breeding from a wild progenitor to agricultural variety are among the more familiar forms biotechnologies have assumed, as is the development in this century of hybrid corn. Taken together, these techniques have resulted in manyfold increases in yields in food crops.[8] Not only does the human species rely on these "mutualists," the organisms that have evolved with or through us; it seems equally true that we cannot survive without the wilds from which we and they have emerged.[9] Many environmental protection issues, including aspects of concerns related to recombinant DNA (rDNA), emerge from the competition between human endeavors and the very environments from which all emerges. We have come far, some would say too far, from the ability simply to cultivate. Now we can engineer and even create.[10]

The history of genetics is illustrative of the scientific revolution and itself the source of profound questions. The chronology is nothing less than astonishing:[11]

1865	Mendel describes the principles of inheritance, "Mendelian genetics."
1953	Double-helix structure of DNA, the basic molecule of inheritance, is revealed.
1957	DNA replication through "unzipping" and exact duplication of resultant strands revealed.
1961	Basic chemical structure (order of amino acids) of proteins found to be coded by triplets of biochemical bases comprising DNA molecules.
1960s	Messenger RNA isolated; other forms of RNA found; all related to the process of translating DNA in cellular nucleus into proteins throughout the cell.
1969	First isolation of a pure gene from bacteria occurs.
1970s	Mapping of genes on chromosomes commenced.
1970	A yeast gene synthesized in laboratory.

1971	A bacterial gene introduced into a human cell with retention of gene function.
1973	A chimera first created through genetic engineering at the molecular level.
	First purification of a human gene.
1976	A human gene synthesized from nucleotides.
1977	System of chemical reactions devised to interrupt DNA molecule at specific locations.
	Original DNA reproduced from RNA.
1981	Gene successfully transferred between animal species.
	First computerization from genetic code of gene synthesis.

One estimate puts contemporary doubling of biological knowledge at a rate of five years and that of genetics at 24 months.[12] With microorganisms we have come to the point that a species can be found to thrive on almost any medium—to form products or to "eat" pollutants such as oil—and can be altered genetically for maximizing the process.[13]

To summarize, before embarking on an examination of "present history," we have discovered that DNA is in effect a set of plans for each protein, the cellular workhorses ultimately serving the entire organism from conception to demise. Proteins are comprised of subunits, the amino acids. Each amino acid is identified throughout biological species by a specific codon, a triplet of nucleotides, each a chemical unit of the DNA molecule.[14]

Present History: Molecular Genetics and Recombinant DNA

> "The thought of recombining living material into an infinite number of new combinations is so extraordinary that the human imagination is barely able to grasp the immensity of the transition at hand. . . . Our world . . . is forged in fire."[15]

The human species has used science to progress from pyrotechnology to biotechnology.[16] Today's biotechnologies include not only rDNA processes but also cell fusion and bioprocessing techniques that supersede traditional techniques.[17] Most observers, however, mean rDNA when they speak of biotechnology in the singular. For all the associated rhetoric, the process is no radical departure from more classical genetics. Instead rDNA represents an acceleration and expansion of preexisting genetic practices.[18]

Recombinant DNA is not threatening in and of itself. It happens all the time in the course of sexual reproduction, a process ubiquitous from bacteria to humans. (It is worth noting that at one point during the rDNA controversy human sex did meet the current definition of moderate risk if conducted in the laboratory.)[19]

The science of rDNA utilizes ("restriction") enzymes to "snip" the DNA molecule at selected points and other enzymes ("ligases") to rejoin the molecule in a preselected order. In recent years methods have been constantly improved for actually extracting and purifying specific genes on the DNA molecule.[20] This laboratory procedure mimics conventional exchanges of genetic material and naturally occurring mutation. Restriction enzymes and ligases, the scissors and glue of rDNA, are essential to those natural processes.[21]

There are certain basics in rDNA processes. A suitable vector, a carrier, such as a plasmid, a virus, or a constructed hybrid of the two is required. This vector must be taken up by the host organism and be capable of independent replication within the host. That replication must be of the donor material carried by the vector. Then there must be a system for distinguishing those cells carrying donor DNA from those carrying only that of the host. Manipulations of these precise molecular structures is so sophisticated as to the point of incredulity among most nonscientists.[22]

Knowledge of molecular genetics had in fact outgrown its disciplinary boundaries as early as 1984. As a result, genetics has served to integrate biochemistry, molecular biology, and cell biology. The findings among the disciplines are necessarily being correlated and integrated with a consequent understanding of life that would have been incomprehensible only a few years earlier.[23] In the mid-eighties two general scientific questions remained open along with a myriad of more precise questions incorporated by the generalities. The two questions were, how genes are controlled during embryogenesis, the development of the individual from conception to birth, and how genetic components recombine from a general cellular product to one fitted more precisely to a particular organism or physiological process.[24]

The areas in which rDNA has been anticipated to be of service and in which the technology has already been effective are truly astounding. They range in application from medicine to agriculture to environmental protection. This molecular tool is capable of syntheses, transformations, degradations, and concentrations with a precision seldom met elsewhere in either chemical or biological engineering. Recombinant DNA may only be one form of molecular engineering, but it is the most powerful.[25] It is expected to meet major industrial needs by the year 2000.[26] Computer processing of information has provided an organizational mechanism allowing the manipulation of the huge quantities of molecular information essential to rDNA techniques.[27] Other than the computer, rDNA's requirements are surprisingly modest.[28]

Ambitions for rDNA are anything but modest, and many are already being realized. These ambitions are directed to the diagnosis, prevention, and control of human and animal diseases, to nutrition and promotion of growth, and to genetic improvement of animal breeds. They are found in the enhancement of certain plant characteristics and in the genetic manipulation of microorganisms for a variety of purposes. Along the way to commercial products these techniques also expand fundamental knowledge in all the areas of interest.[29] There are, however, limits.

There is no attempt to achieve the unlikely goal of "engineering" whole organisms, even as relatively simple as a bacterium. Only a few characteristics can be manipulated at best.[30] Nevertheless, as early as the mid-seventies, small entrepreneurial firms were being formed to compete with an international establishment of large firms exploring rDNA possibilities.[31]

That there might be hazards with this as with all technologies cannot be denied. That some recombinants, like pathogenic organisms, could be dangerous was virtually certain. But scientists are optimistic. Knowledge increases understanding. Less than ten years into the modern history of rDNA found risks being characterized as acceptable whereas previously they might have been unimaginable horrors.[32]

For all the promise in medicine, rDNA has been predicted to attain its greatest promise in agricultural applications.[33] It is here, with potential releases to the environment, deliberate as well as accidental, that politics become a potent force affecting applications of this bioengineering innovation. Where some saw promise, others saw a specter of "unfathomable" consequences, more threat than promise.[34] With such controversy and politics, regulation is sure to follow.

In fact, political response to rDNA parallels the preceding scientific and technological chronology:[35]

1973	Gordon Conference: First discussion of potential for good and for harm.
	Maxine Singer letter to *Science* and National Academy of Science ("NAS") and additional communications raising issues of safety and the wisdom of proceeding with research.
	Call for voluntary pause in certain kinds of research.
1974	NAS correspondence to *Science* recommending management approaches; resulted in Asilomar and bureaucratic programs for that purpose.
1975	Asilomar Conference: Precedent-setting consultation within scientific community to consider propriety of proceeding with rDNA research programs.
1976	Formal NIH guidelines for research issued.
	Genentech, a major rDNA entrepreneur, founded.

1977 Letter from scientist and former critic of rDNA techniques to National Institutes of Health (NIH) concedes no danger associated with using a common human bacterium as a vector.

1980 U.S. Supreme Court accepts patentability of engineered organisms, the first time "life" is found to be patentable.[36]

VOICES OF THE PUBLIC

> "*Debate*... has always meant something well ordered and ostensibly sensibly rational, where all parties strive to be well informed, logical, and with luck, unemotional. But the highly contentious melodrama about the usage of gene-splicing methods at its peak... was anything but that."[37]

Throughout the history of science, there has been the fear of emerging knowledge. In part, it is the fear of the unknown, a sense that scientists are probing where humans have no business. Human societies experience a common tension between the comfort of tradition and the compulsion to explore uncharted territory.[38] Notably, in the case of rDNA, it was the prudence of the scientific community rather than fear in the public that opened the door to debate.

Voices of the Scientific Communities

> "Involved scientists recognized a problem, took an initiative directed to the public interest, devised an interim solution, and succeeded in defending it against significant opposition that would have imposed far greater restriction."[39]

Asilomar stands representative of the voices of scientists in the matter of rDNA hazards and their control. The strategy of this conference of interested scientists was to classify risks associated with all conceivable recombinants in terms of probability and mitigation. The scientific community anticipated guidelines to be administered by NIH.[40] Most of the responsible community were calling for prudence, a balance between foolish chances and unnecessarily restrictive controls. Significantly, the fact remains that there has been no documented case of accident or injury attributable to processes of rDNA. In fact, the matter has pretty much turned invisible to the public at large within the decade.[41] Nevertheless, Asilomar proved to be the first step in the public policy process, still evolving,[42] with or without public participation.

Asilomar consisted of 150 molecular biologists and geneticists. The only nonscientific representation was through lawyers and journalists. The conference was indicative of professional recognition that the interests went beyond science and that society at large needed to be kept informed. The task at hand was to devise a rational and effective form of management for the perceived risks while assuring continuation of research.[43]

Discussion was not as simple as might be expected. No one could predict the type of harm reasonably expected or quantify the probabilities of any risks. Dissenting scientists held the advantage. Unlike rDNA's proponents, opponents were limited only by their imaginations.[44] One professor was implacable in his opposition, remaining unconvinced by attempts to demonstrate safety and concerned about the entry of industry into the field.[45]

Dissidents have not been above framing questions designed to frustrate answers. Not only could such questions be highly compounded and loaded with assumptions and allegations; they could be further loaded with "tonal allegations." References were made to Dr. Frankenstein, to playing God, and to crossing species boundaries with an implication of exceeding natural law.[46]

The division, of course, spread. As the issues continue to arise for consideration, scientists are motivated by their implicit moral contract with society to do no harm, while preserving their autonomy in the name of vigorous inquiry. Nonscientists, however, find their values and way of life pressured by seemingly uncontrolled science and technology. Nonscientists strive to retain a voice in choosing among the options opened to them.[47] Defeating early communication is the conviction of most scientists that they risk their reputations if they indulge in creative thinking with the public as audience. Thus, overly intrusive public scrutiny can be truly inhibiting.[48]

Aside from one public forum with its exchange between scientists and the public,[49] it is legitimate to query the extent to which scientists have engaged in educational efforts.[50] But before turning from scientists to the public at large, attention to the marketplace is informative.

Voices from the Marketplace

> "Biotechnology has been oversold as a way of making money. Commercial time scales have been unrealistic. . . . Overall the technology sparkles, but it is far from clear who most stands to gain what, and when."[51]

The markets most likely to be affected by rDNA are wide-ranging. The most significant to the present discussion are health care and pharmaceuticals, a variety of chemical manufacturing processes, and the management of pollutants and toxic wastes.[52] Artificial enzymes are the chemist's and the pharmacologist's dream. That they are also the dream of entrepreneurs large and small is reflected in the array of firms interested in developing rDNA for their own purposes.[53]

Controversy has been intense with reference to the partnership between industry and academia, particularly appropriate in developing rDNA technology.[54] Although the partnership was hardly new, molecular biology was a new element bearing many of the conflicts that were to be voiced by the public at large. Much of the controversy was founded in large part in ignorance.[55] In addition, there were questions of complications surrounding free speech and free access to information underlying the openness requisite to scientific debate.[56] Later, when early field tests led to even more controversy, interested firms found themselves seeking to lay the groundwork to reshape public perceptions,[57] somewhat belatedly. In many cases firms do engage in careful planning for public relations through a variety of techniques.[58]

Voices for Sectors of the Public

> "It is difficult to take demands for bans on 'all' recombinant DNA research seriously when they come from those who demand to be heard but do not stay to listen. . . . The cautious analytical approach is a discouraging tactic against fear, mysticism and political opportunism."[59]

Both sides of the rDNA debate have been attended by hyperbole. Unlimited possibilities were countered by extreme uncertainty, giving rise to hope on one hand and fear on the other.[60] The mixed reaction is not tempered when public perceptions of science, scientific process, and scientists are at best limited and at worst greatly distorted.[61] The sensationalism surrounding introductions of rDNA technology to the public only exacerbated the ignorance and suspicion. Images from the entertainment media hovered ominously.[62] Still, it is the aura of drama and the unusual that seems prerequisite to gaining public attention at all.[63]

Journalists, even those with scientific credentials, have no easy task in conveying esoteric innovations in a style both understandable and readable. Some reporters succumb to emotional and melodramatic presentations because these are better "copy" than a more responsible presentation. They are also highly effective in moving the public.[64] Further alienation of the public can be attributed to envy of those research scientists reportedly turned entrepreneur and becoming wealthy in the process.[65]

Whatever the source of information and motivation, public perceptions are among the ten factors the congressional Office of Technology Assessment analyzed in addressing our future competitive position in this exploding field of endeavor.[66] The argument is justifiably raised that all who might be victims of a program of development from novel technology should have a voice, despite the fact that risk is a part of life and one cost of gaining new knowledge. In the case of rDNA the further argument can be raised that all people everywhere should be informed so that they might take a stand. At the minimum, those in the immediately affected community must be informed and heard.[67]

Asilomar set the stage. Over the next several years NIH took the lead among federal agencies in establishing control of rDNA research and marketplace appearances. Still, because no forum for concerted review was ever established after Asilomar, public impressions continued to reflect worst-case scenarios instead of a more realistic image. Asilomar did, however, serve to humanize the public image of scientists. What could have been a perception of a brilliant conspiracy became one of a conscientious effort on the part of people of good intentions.[68] While paranoia can occasionally crop up, the press seem to be taking a reassuring posture with a notably beneficial influence on the public.[69]

Whatever one's perspective, it seems inevitable that rDNA and other bioengineering techniques place our species at a crossroads in our technological history.[70] It is only fitting that some guideposts be put in place.

NOTES

1. B. ZIMMERMAN, BIOFUTURE: CONFRONTING THE GENETIC ERA (1984), at 72.
2. E. ANTEBI and D. FISHLOCK, BIOTECHNOLOGY: STRATEGIES FOR LIFE (1985), *passim*. This volume, a "coffee table" version of these sciences, offers appendices of definitions and of biotechnology policy and its regulation by the U.S. Department of Agriculture. The text itself provides a thoroughgoing portrait of the sciences, technologies, and personalities in genetics and genetic engineering, accompanied by extensive graphics.
3. Ibid.
4. Janzen, 236 SCI., Jun. 5, 1987 letter, at 1159.
5. Antebi and Fishlock, *supra* note 2.
6. Zimmerman, *supra* note 1, at 71.
7. Ibid., at 179.
8. Harsanyi, *Beyond Recombinant DNA: The Next Twenty Years*, in THE GENE-SPLICING WARS: REFLECTIONS ON THE RECOMBINANT DNA CONTROVERSY (R. Zilinskas and B. Zimmerman, eds., 1986), at 207. All the contributors to this anthology were involved in the rDNA controversy of the seventies and early eighties.
9. Janzen, *supra* note 4.
10. J. RIFKIN, ALGENY (1983), at ix.
11. The chronology is combined from Antebi and Fishlock, *supra* note 2, at 226, and Rifkin, ibid., at 8–10. *See also* A. STURTEVANT, A HISTORY OF GENETICS (1965) and CLASSIC PAPERS IN GENETICS (J. Peters, ed., 1959) for the history of "classical" genetics. For more recent advances, *see* BIOTECHNOLOGY & BIOLOGICAL FRONTIERS (P. Abelson, ed., 1984), and BIOTECHNOLOGY: THE RENEWABLE FRONTIER (D. Koshland, ed., 1986).
12. Rifkin, *supra* note 11, at 7.
13. Harsanyi, *supra* note 8, at 207.

14. J. CHERFAS, MAN-MADE LIFE: AN OVERVIEW OF THE SCIENCE, TECHNOLOGY AND COMMERCE OF GENETIC ENGINEERING (1982), at 3–25 *passim*. A nucleotide is a deceptively simple combination of small molecular components—a sugar, a phosphoric group, and one of four nitrogenous bases.

15. Rifkin, *supra* note 10, at 14. Jeremy Rifkin, no friend of rDNA, is nevertheless capable of reducing the procedure to a single, extraordinarily clear paragraph. Ibid., at 7.

16. Ibid., at 6. For a history of the sciences behind rDNA technology and the status of technology in 1984, *see* Zimmerman, *supra* note 1, Appendix, at 249–279.

17. Antebi and Fishlock, *supra* note 2, at 6–9.

18. Davis, *Paranoia Invades Report on Man-Made Microbes*, WALL ST. J., June 16, 1988, at 22. Dr. Bernard D. Davis was a professor emeritus of bacterial physiology at Harvard Medical School.

19. Cherfas, *supra* note 14, at 137.

20. Ibid., at 24–25.

21. Antebi and Fishlock, *supra* note 2, at 6, 53 and 54.

22. COMMERCIAL BIOTECHNOLOGY: AN INTERNATIONAL ANALYSIS (U.S. Congress, Office of Technology Assessment, 1984).

23. Zimmerman (1984), *supra* note 1, at 46.

24. Ibid., at 230, 231–233 *passim*.

25. Harsanyi, *supra* note 8, at 207.

26. Ibid., at 209.

27. Rifkin, *supra* note 10, at 21.

28. Newell, *The Science and Technology of Gene Splicing*, in Zilinskas and Zimmerman, *supra* note 8, Appendix at 248.

29. COMMERCIAL BIOTECHNOLOGY, *supra* note 22 *passim*. *See also* Antebi and Fishlock, *supra* note 2, at 133–135 and *passim*. Advances are routinely reported in *Science* and the *Wall Street Journal* as well as in the broader spectrum of mass media. *See*, e.g., Wells, *A Real Gas of a Gene Could Put the Bite on Any Bad Apples*, WALL ST. J., Sep. 1, 1988, at 4.

30. Brill, 236 SCI. 1159, Jun. 5, 1987 letter.

31. Antebi and Fishlock, *supra* note 2, *passim*.

32. Cherfas, *supra* note 14, at 138.

33. Ibid., at 235. *But see* Moses and Hess, *Getting Biotech into the Field*, Fall ISSUES IN SCIENCE AND TECHNOLOGY 35 (1987).

34. Rifkin, *supra* note 10, at 15.

35. Chronology combined from COMMERCIAL BIOTECHNOLOGY, *supra* note 17, at 489–497, *passim*, Cherfas, *supra* note 14, at 129–134, and Grobstein, *Asilomar and the Formation of Public Policy*, and Zimmerman, *Science and Politics: DNA Comes to Washington*, in GENE-SPLICING WARS, *supra* note 8, 3, and 33, respectively.

36. Diamond v. Chakrabarty, 447 U.S. 303, 100 S.Ct. 2204 (1980). Although this decision addressed not molecular biotechnologies but the crossing of bacterial strains by more traditional methods, the Supreme Court opened the door for patents in this more innovative technology. *See* Antebi and Fishlock, *supra* note 2, at 187–189.

37. Zimmerman, *supra* note 1, at 141.

38. Ibid., at 199.

39. Grobstein, *supra* note 35, at 6.

40. Ibid., at 5.
41. Zimmerman, *supra* note 1, at 142–143, 145.
42. Grobstein, *supra* note 35, at 4.
43. Ibid.
44. Harsanyi, *supra* note 8, at 219.
45. Cherfas, *supra* note 14, at 141.
46. Schmeck, *Recombinant DNA Controversy: The Right to Know—and to Worry*, in GENE-SPLICING WARS, *supra* note 8, 93, at 116. *See also* Cherfas, *supra* note 14, at viii.
47. Grobstein, *supra* note 35, at 8.
48. Zimmerman, *supra* note 1, at 14.
49. Singer, *The Recombinant DNA Debate*, 196 SCI. 127, Apr. 8, 1977 editorial. This is the Maxine Singer who "went public" early in the course of rDNA research. She and Dieter Soll expressed the consensus of the Gordon Conference on Nucleic Acids in a September 21, 1973, letter to *Science*, thus setting the stage for Asilomar. *See* e.g., Zimmerman (1986), *Science and Politics: DNA Comes to Washington*, in GENE-SPLICING WARS, *supra*, note 8, 34.
50. Nader, *Technology and Democratic Control: The Case of Recombinant DNA*, in GENE-SPLICING WARS, *supra* note 8, 139, at 155. Ms. Nader asserts that conferees at Asilomar sought to alert their colleagues rather than to educate the public. Ibid.
51. Langley, *Corporate Strategies for Diversification*, in Antebi and Fishlock, *supra* note 2, 204.
52. Harsanyi, *supra* note 8, *passim.*
53. Antebi and Fishlock, *supra* note 2, *passim.*
54. COMMERCIAL BIOTECHNOLOGY, *supra* note 17, at 411–421 *passim.*
55. Cherfas, *supra* note 14, at 200–206, *passim.*
56. Nader, *supra* note 50, at 147. Legally, this complication relates to intellectual properties, primarily patents and trade secrets.
57. Sun, *Preparing the Ground for Biotech Tests*, 242 SCI. 503, Oct. 28, 1988 News & Comment.
58. Ibid., at 504.
59. Singer, *supra* note 49.
60. Harsanyi, *supra* note 8, at 206.
61. *See* e.g., Zimmerman, *supra* note 1, at 1–2 and *passim*, and COMMERCIAL BIOTECHNOLOGY, *supra* note 17, at 489.
62. Ibid., at 38. *See also* Harsanyi and GENE-SPLICING WARS, *supra* note 8, at 218 and xi, respectively.
63. Ibid., at 4.
64. Ibid., at 105–106 and *passim*. *See also* Schmeck, *supra* note 46, *passim.*
65. COMMERCIAL BIOTECHNOLOGY, *supra* note 17, at 491.
66. Ibid., at 9 and 265. This report chronicles alterations in pertinent perceptions among the public from 1977 through the early eighties. Ibid., at 466–497 *passim*. In 1987 more than 80 percent of the public reportedly supported field experiments with rDNA organisms in the belief that benefits would outweigh risks. More than half were willing to chance limited ecological damage, as long as the potential risk is known. Among those evaluating risks, university scientists were most trusted, fol-

lowed by public health officials and representatives of federal agencies. Crawford, *Mixed Views on Biotech*, 236 SCI. 1179, Jun. 5, 1987.

Another survey of public attitudes was reported in 1991. This time scientists engaged in rDNA research were asked how litigation and public advocacy had affected their work. The results were mixed; some thinking the involvement beneficial, some harmful, and others with mixed reactions. Many were greatly concerned about the effect on future direction, including international competitiveness. Briefings, 251 SCI. 871, Feb. 22, 1991.

67. Michael, *Who Decides Who Decides? Some Dilemmas and Other Hopes*, in GENE-SPLICING WARS, *supra* note 8, at 121–137 *passim*.

68. Grobstein, *supra* note 35, at 7.

69. Davis, *supra* note 18.

70. Rifkin, *supra* note 10, at ix and 6.

CHAPTER 8

Regulation, Voices of the Public, and Products of Recombinant DNA: The Intersection with Policy and Law

> "While the scientists at Asilomar could perceive in the National Institutes of Health (NIH) a nurturing hand both familiar and friendly, the federal government as a whole, with its plenary responsibility for public policy, was only dimly perceived. For the most part government was viewed askance as a threatening image, and its activation was to be avoided if at all possible."[1]

The federal government does show something of a split personality in its relationship to science and technology. Innovations are advanced under constitutional mandate through the patent system and grant programs. But those same innovations may also be subjected to control through statutes and regulations.

ADVANCEMENT OF THE USEFUL ARTS AND SCIENCES

> "Recombinant DNA technology faces our society with problems unprecedented not only in the history of science, but of life on Earth. It places in human hands the capacity to redesign living organisms, the products of some three billion years of evolution."[2]

In the case of molecular biotechnologies the federal government had been expending millions of dollars on an annual basis, even before applied research came to the fore.[3] In 1980 the first process patent for "biologically functional molecular chimeras" was granted.[4] By 1988 there was an international competition to achieve excellence in biotechnology in the markets identified in the last chapter. Here in the United States, funds were directed primarily to biomedical research, with less going to plant biology, where major firms are active. To a certain extent, however, progress was being impeded by regulation and other legal actions.[5] There were calls for greater investment in these technologies.[6]

In brief, the government can facilitate or impede commercialization of rDNA techniques and products through the array of laws within its aegis. Relative freedom and rapid capitalization provide advantages. But funding for basic research and personnel training was a perceived need in the mid-eighties.[7] At that time the public actually had little to say about directions of rDNA research. Apparently, reliance was placed in the marketplace. It must be noted, however, that there are few institutions or advocacy networks for the public to make its voice heard for purposes of directing any given technology.[8] This has been the case since the days of nuclear energy controversy.[9] As Burke Zimmerman put it later, "The power of technology is now so great that many believe its directions cannot be left to chance, or to free enterprise. . . . What is needed is the development of an ethic to govern technology, or at least to establish a process by which ethical considerations are taken into account."[10]

THE CALL FOR CONTROL OF RECOMBINANT DNA TECHNOLOGIES

> "[R]egulations not based on relevant scientific principles but, rather, on alarmist views, will only hinder research and development and U.S. competitiveness in projects that are expected to help agriculture and the environment."[11]

No one can doubt that health, safety, and environmental protection are legitimate concerns whenever a new technology is developed and begins to enter the marketplace. Recombinant DNA is no exception.[12] Still, its risks were for a time subjected to gross overstatement, an observation supported by NIH's subsequent move to relax its control through rDNA research guidelines.[13]

The history intervening between Asilomar and eventual relaxation of NIH control goes beyond merely interesting. It provides a valuable lesson in the interaction between science and law that may well be prophetic.

Scrutiny of rDNA began at the instigation of the potentially regulated community itself. As early as 1975 formal inquiry included scientists through peer review and oversight functions. Later, representatives of public health, law, consumer affairs, and public policy were included.[14]

In July of 1976 NIH issued its guidelines in recognition of a possible risk of undue pathenogenicity or adverse environmental effect should an rDNA organism be accidentally released.[15] The preamble recites relevant history and notes that the NIH guidelines would replace the Asilomar recommendations. These guidelines, prohibiting deliberate releases and providing extreme specificity for technical procedures, would be widely disseminated.[16]

By the end of the year advocacy groups had petitioned NIH for a public hearing on the matter of rDNA control in order to open debate to the public and expand the issues before NIH. Other petitions followed, and a coalition was formed among specialists and concerned citizens.[17] A draft revision was published in September.[18]

Separately from the actions within the scientific community and through NIH, an interagency committee was formed late in 1976. This cooperative effort quickly established that new law was required to assure compliance with the NIH guidelines in a consistent manner.[19] Theirs was not the only voice calling for more complete control than NIH was offering.

Communities fearful of local enclaves of biotechnologists were also debating the issues as they perceived them. Notable among these were Cambridge, Massachusetts, Ann Arbor, Michigan, and the state of New York.[20] By 1981 at least six communities had enacted ordinances addressing rDNA companies. Four of these were in Massachusetts.[21] Complicating the regulatory issues and underlying technology before these regulators were ethical issues. Only one community directly addressed ethical concerns, with a prohibition of experiments involving humans.[22] The matter of ethics obviously adds another level of complexity to an already baffling matter.[23]

The most prominent among these local efforts is almost certainly that of Cambridge, where the mayor and his advisor, Nobel laureate geneticist George Wald, headed up a coalition of scientists and citizens forming the Cambridge Experimentation Review Board. Their target: Harvard's rDNA research. The professed fear centered on the escape of genetically modified microorganisms to wreak plague and pestilence on Cambridge and beyond. In the end even this coalition accepted the NIH guidelines as the best approach. Cambridge merely added provisions, including stricter containment.[24]

The Cambridge experience, for all its reported dramatics and aura of panic, provides some valuable lessons for future intersections of science and the politicolegal system. In the end this coalition of interests affected by this innovative technology endorsed the regulatory mechanism established by the proponents of the technology and its most knowledgeable regulator. Cambridge also demonstrates that the involvement of the lay public need not result in some mindless disaster of excessive regulation. At the same time, however, the process was in a sense burdensome and expensive and arguably unnecessary.[25] But just this sort of public involvement is contemplated in the basic structure of our economic and political system. It serves as an educational tool, one which those charged with education should heed rather than scorn.

Recombinant DNA did not escape the interest of Congress. At the close of 1976 the House Committee of Science and Technology received a report chronicling the evolution of the technological issues.[26] That the issues went beyond the matter of rDNA was made abundantly clear. Recognizing that highly significant scientific issues would continue to arise nationally, as had been the case for a decade, the authors of the report acknowledged that the rDNA controversy was focusing attention on the ubiquitous need for evaluation of societal effects of any innovative research. In the process, the task of integrating contrasting perspectives would only assume increasing urgency. In the case of rDNA and, by projection, comparable future innovations, the rewards would be great, but the risks would be no more than partially defined.[27]

Before the decade ended, competing proposals were before both houses of Congress. It is telling that the scientists were outraged by the best of the legislation while the marketplace remained calm. In a profound (but hardly unique) naivete, the scientists were predicting the end of open scientific research at the hands of "mindless bureaucratic oppression."[28] Then, when a so-called nonfiction work on the cloning of a human being made headlines, Congress—and science—was in an uproar. There was, nevertheless, a benefit to be extracted: The foremost scientists in the field came to Washington to educate Congress and, not incidentally, the public in the sciences of cell biology. As a result a record was made of this science in terms accessible to the layman.[29]

There were the inevitable legislative proposals, but none of them were as extreme as the scientific community had fearfully predicted. In the end rDNA and bioengineering became something of a nonissue.[30] The underlying reasons for the defeat of a legislative proposal will never be entirely clear. The matter of biotechnology control is no exception. The cynical can identify with the observation that a bill subjected to lengthy and intensive debate is manipulated to the point of no return. In colorful biological language, "the more embellished and grotesque it becomes, until . . . its original purpose is lost. Like a dinosaur overspecialized for a specific ecological niche, it loses its . . . adaptability, becoming a ponderous version of what it once was."[31]

In essence, the initial fears dwindled, perhaps in response to the efforts of the scientific community, through the NIH guidelines[32] and, almost certainly, because no incidents drew the fevered attention of critics or the press. In addition to the lack of support for the early scenarios of disaster, formal risk assessment actually downgraded the probability of harm, and molecular biologists themselves were gaining public confidence through bringing members of the public as well as additional experts into the ongoing policy formulation process.[33]

For all the seeming intelligence behind the ultimate congressional quiescence regarding biotechnology, even Burke Zimmerman later admitted that an element of the scientific community maintained a kind of aloof disdain, not even bothering to distinguish the actual proposals or to become familiar with the regulatory process.[34]

At this juncture it is worth noting a point of biological interest. The field is microbial ecology. As late as 1984 it could be asserted with some veracity that the ecology of soil bacteria had been subjected to extensive study. There was no scientific evidence addressing potential perturbations of soil microcosms caused by rDNA bacteria intended for agricultural improvements. Would the introduced bacteria displace their natural counterparts and disrupt existing ecological balances?

A serious point is raised here, even where most soil biologists might concur that a perturbation is unlikely.[35] Certainly, some element of governmental control was in order, then and now.

BUREAUCRATIC RESPONSE AMONG PROTECTORS OF ENVIRONMENTS

> "The response to a cornucopia of conjectured risks of gene splicing has not been trivial. New relationships have been created between science and its social institutions. These relationships are not only unique to the field of biology, but some are unprecedented for the entire enterprise of science."[36]

A number of federal executive agencies are charged with the protection of a variety of environments, from the workplace to rural settings to local and national environments and now extending to the biosphere itself.[37] One point of departure for these agencies, and the point of departure for environmental issues surrounding rDNA and its products, is the Environmental Impact Statement, mandated by the National Environmental Policy Act (NEPA).[38] Whenever there is a major federal action that may significantly affect the human environment, the responsible agency must seriously consider that potential affect.[39] Although a federal agency itself may not engage in rDNA research or production, it may fund such research or license it. Such an act can arguably fall under the NEPA aegis.

From the mid-seventies on, then, rDNA research, development and commercialization proceeded under the benign guidance of NIH. Notably, however, the guidelines were mandatory only for those funded by NIH. Once commercialization commenced, there was no federal regulation. It was then that the Department of Agriculture (USDA), the Food and Drug Administration, and the Environmental Protection Agency took an interest in the proceedings. None of their enabling statutes had been enacted with rDNA in mind. Policies for extending statutory authority, if not mandates, were therefore in order.[40]

At the end of 1984 a federal Coordinated Framework for Regulation of

Biotechnology was proposed. It was adopted in June of 1986.[41] Congressional hearings followed.[42] In addition to the agencies identified above, the Occupational Safety and Health Administration and both NIH and the National Science Foundation were to be involved.[43]

Divisions of labor among the agencies was as might be expected. For environmental protection, USDA[44] and EPA[45] took the lead. EPA would regulate rDNA through the Federal Insecticide, Fungicide, and Rodenticide Act[46] and the Toxic Substance Control Act (TSCA).[47] The underlying premise was that existing law and regulation was generally adequate but that some microbial products would require additional regulatory attention.[48]

Significantly, an additional premise was that biotechnology is not inherently dangerous. Regulatory review, therefore, would be essentially the traditional approach with a possible call for additional data. In fact, the regulators subsequently found themselves unable to define environmental release for purposes of legal control. The approach, then, would be ad hoc.[49] TSCA provided tools intended to develop data for a determination of whether either a new chemical or a new use of a chemical already in commerce required legal control and, if so, whether control would take the form of a negotiated "contract," an order, or an administrative regulation. This is TSCA's Premanufacture Notice process.[50] The first biotechnology Premanufacture Notice was received in 1987.[51] In the same year the first legal outdoor release was authorized. It was reported in a single, brief paragraph in the *Wall Street Journal*.[52]

The philosophy behind field-testing of rDNA organisms, that is, their deliberate release into the environment, consists of predictable elements, two of which are of particular pertinence here: disclosure to politicians, media, and community leaders and establishment of credibility with the local community.[53] One can only hope that such an approach proves to be prophetic at the four-way intersection of science, technology, law, and policy.

REACTIONS AT THE INTERSECTION: MERGER OR COLLISION?

> "The recombinant DNA controversy excited the natural tendency to assume that dilemma[s] involving profound new questions should automatically be exposed to new, untested [political] mechanisms. . . . Fortunately, the maturity [or stubborn tradition] of our political system . . . forced this complex problem to be engaged . . . with the existing framework."[54]

The primary source for the legal status of rDNA processes and products is the judicial system. It is instructive, however, to be advised that the courts required four years to declare the previously reported volume on human cloning to be a fraud.[55]

In 1982 the scene was set for the first encounter between rDNA and the judiciary. An academic scientist sought NIH approval for field tests of a common bacterium altered by rDNA techniques. This "ice-minus" rDNA mutant was intended to reduce frost damage on potato plants. Based on all the data supplied, NIH approved the experiment in 1983, and suit was promptly filed for failure to conduct an environmental assessment in accord with NEPA.[56] The federal District Court in Washington, D.C., enjoined the experiment on the basis that a more thorough environmental assessment was in order. The injunction against the experiment was affirmed on appeal but vacated with regard to other deliberate release experiments.[57]

The same district subsequently dismissed, for lack of case or controversy, an attempt to enjoin the effect of the Coordinated Framework for Biotechnology. The action alleged the document to be so incomplete and inexact as to allow regulation of potentially dangerous products of rDNA to be overly casual. The court found the complaint defective, in part because guidance is not regulation but only a first step in the formulation of relevant policy. In short, plaintiffs were without standing to sue.[58]

In a companion case the same plaintiff was again unsuccessful, this time in an action seeking an order requiring EPA to modify its procedures for authorizing environmental releases of rDNA pesticides. Any injury claimed was held to fall outside judicial cognizance.[59]

While the U.S. Supreme Court has yet to address issues of rDNA, it has taken a stand through the years on innovations in the NEPA context. It has generally held that such questions, often on the frontiers of science, are better decided in legislatures and in administrative agencies charged with authority in point. The judiciary is cautioned to watch procedure and to avoid the merits, with special reference to the acceptability of risks.[60]

Regulatory efforts and judicial response have been closely watched by interested communities and widely reported. In 1980 Maxine Singer observed that since 1950 molecular biologists had enjoyed unprecedented support on the part of enlightened societies and governments. She called the mutual efforts a joint venture and observed that extreme views of the hazards were already deemed groundless.[61] In 1987 an attorney echoed her sentiments with the observation that coordinated efforts had done much to alleviate fears, resolve questions of jurisdiction, and create a uniform approach. Karny did note the continuing need for better definitions,[62] never a matter to be taken lightly.

That the researcher must tread with extreme care was graphically demonstrated in 1988, in the context of rDNA as a weapon against Dutch Elm disease. The scientist in question had inserted rDNA into diseased trees without EPA approval. In the end, there was no violation of applicable guidelines. The resultant lessons were numerous. The scientific community was learning, sometimes the hard way, to fight back against its critics. The National Academy of Sciences (NAS) was suggesting that rDNA represented no unique hazard-and EPA announced its intent to ease its own guidelines in point.[63]

Surprisingly enough, by 1990 the rDNA marketplace was actually lobbying for regulations. The lack of rules for marketing and testing agricultural products was reported to be holding up research and investment. The agencies had yet to agree on issues as basic as the means for determining whether plants or microbes should be subject to special review or exempt from review. In fact, the agencies were on opposite sides of the issues. The regulated community could not make any long-range plans.[64]

The confusion might have been predicted. In 1986, it was observed that, although research agendas and the assessment and control of technological impacts have their governmental institutions, no single institution has the authority to set and establish priorities for the placement of rDNA in clinical, industrial, or agricultural applications.[65]

CONSOLIDATION AND SYNTHESIS

> "The arguments on all sides became more intense, as 'educated' assessments of... risk collided, and, because the facts were sparse, could not be resolved. Many... scientists jumped into the fray, and the polarization of the scientific community... and other groups in our society, was well underway."[66]

The rDNA debate stands as a case study regarding nationwide science policy formulation. The events commencing with Asilomar chronicle the ways in which a vital scientific issue can simultaneously influence practitioners in the affected discipline and national science policy. Obviously, if the public is to assume a meaningful role in the process, a clear statement of the issues must be proferred along with an evaluation by technical experts of the potential for benefit and harm.[67]

The public must be afforded that meaningful role, but the rDNA debate is the modern harbinger of future controversies arising from tension between the imperative for technological advancement and national values.[68] With future advances, as was the case with rDNA, the perception may be one of catastrophic consequences. Here two extremes may be offered: zero risk versus any and all risk for the sake of advancement. Society is unlikely to accept either. The acceptable balance, however, can be predicted to be a major source of controversy. At this juncture a fundamental disagreement remains: Where is the line to be drawn? Just when someone essays an answer, the line moves.[69]

In the case of rDNA, the community of scientists opened the way to public debate. Asilomar is unique, and it was timely. But, as early as 1973, a step oriented more to public information was established by the American

Society for Microbiology with its establishment of a permanent Public Affairs Committee. This group was charged with establishing the means of recognizing and responding to research developments likely to invoke policy formulation. The move was not unique among scientific associations.[70] In 1977 participants in an NAS forum took the position that such organizations have an obligation to inform and to speak out on scientific issues. According to NAS a major biohazard is ignorance. After all, public protection depends ultimately on knowledge and the exercise of judgment. NAS and others set out to practice what was being preached.[71]

Public education in the sciences need not be limited to the grand scale represented by NAS and national scientific societies. In 1979 the University of Michigan at Ann Arbor conducted a series of lectures and discussions open to the public. The series addressed not only the biological background of rDNA but also the associated ethics and the legal and sociopolitical issues. The series met with mixed reviews from within and beyond the scientific community.[72] Still, this was a step in the right direction.

It is only logical to assert that, if scientists have an obligation to inform, then the public has the corresponding obligation to become informed. In point of fact, there is considerable logic to the assessment that the most important element of the rDNA controversy has been how the technology is being perceived by the public. At least in the early years, rDNA provided an exercise in human imagination,[73] largely unfettered by reality.

Despite Asilomar, it took a year for the controversy to reach the general public. This is an unfortunate circumstance, because, for better or worse from a variety of vantages, public perceptions will strongly influence the progress of research, industrial development and, quite rightly, the related safety of society. After all, it stands to reason that the better informed a person is, the more rational that person's responses will be. Unfortunately, though the public may have an obligation to become informed, this nation affords no means to compel the individual to read—especially beyond the headlines,[74] sensible or sensational. More realistically, we have yet to develop an identifiable institution charged with the requisite authority for setting appropriate priorities, and there remains little precedent to do so in the politicolegal system. There simply is no overseer.[75]

What, then, *are* the governmental duties corresponding to those of science and society? Congress did not rise to the challenge provoked by rDNA. Arguably, that institution is not equipped to handle scientific questions (although it probably should assume that obligation), but it is charged with the mission of establishing political and regulatory mechanisms demanded for assessing technological advances. Instead, along with all the rest, Congress shifted with the wind and allowed the novelty to wear off, more out of inertia than anything more responsive.[76] The question, then, remains open: Who *does* decide who decides? The answer promises to remain indefinitely problematic.[77] Inertia of our institutions is not the proper response, nor is it acceptable.

Clearly, the rDNA controversy has afforded us a number of lessons, of undeniable future import. The lessons suggest yet again that where science and technology are meeting the politicolegal system, the public is in effect disenfranchised. Reenfranchisement demands that we take the lessons seriously and act upon them.

The lessons can be consolidated as four sources of discontent, together a reflection of the complexity behind the demands for policy formulation at the intersection: (1) the ideological and psychological virtues of citizen participation bringing to the fore what may be an incompatible mixture of ethical issues; (2) public funding of research, in company with other circumstances, leading to a growing challenge to the autonomy of scientific research; (3) widespread questioning of the legitimacy of existing institutions, with regard to their decisionmaking authority and their competency—coupled with a distrust of large organizations; and (4) widening recognition that neither scientists nor the applications of their science are value-free.[78]

An overriding point to ponder is why policymakers must commence deliberations, as they did with genetic engineering, with all-out formal hearings addressing what had to have been obvious scare tactics on the part of the ignorant and the manipulative. Why would they not first assess the quality of the sources? In today's era of ongoing scientific revolution these questions are more than rhetorical.

A reality here is that we must learn to accept uncertainty and to live with it without paralysis.[79] The Cambridge experience, where the local public addressed their fears with scientists, taught that the lay public *can* learn scientific esoterica to the end of reaching rational judgments affecting its entry into society.[80] We have also been taught that minority positions among scientists must be presented, but at least the relative size of that minority must also be noted.[81]

The pessimistic observation from the rDNA controversy is that the public possesses excessive expectations that its institutions will provide a balanced response to innovation. There is serious doubt that scientific institutions are capable of evaluating the facts or accurately presenting them to society at large.[82] Certainly, the parallel congressional response, however perceived in retrospect, also reveals a real need for greater communication across a variety of lines.[83]

The optimistic observation is to the effect that our institutions are possessed with a real genius for allowing progress in face of confrontation. Our government *can* serve the public interest without impeding the scientific endeavor. With rDNA we did discover formal regulations to be overly rigid and immutable and sought out effective alternatives.[84] We must continue on a course of appreciation for the wonders of science and technology, but from a realistic perspective. We have a shared responsibility to make wise use of all we learn.[85]

NOTES

1. Grobstein, *Asilomar and the Formation of Public Policy*, in THE GENE-SPLICING WARS: REFLECTIONS ON THE RECOMBINANT DNA CONTROVERSY (R. Zilinskas and B. Zimmerman, eds., 1986), 3, at 5. In 1986 Clifford Grobstein was Professor of Biological Sciences and Public Policy at the University of California (San Diego).

2. G. Wald, *The Case against Genetic Engineering*, in GENETIC ENGINEERING, HUMAN GENETICS, AND CELL BIOLOGY: EVOLUTION OF TECHNOLOGICAL ISSUES, December 1976 Supplement Report II on DNA RECOMBINANT MOLECULE RESEARCH, prepared for the Subcommittee on Science, Research and Technology of the House Committee on Science and Technology, Appendix 12, at 239. George Wald received the Nobel Prize for his pioneering work in genetics.

3. COMMERCIAL BIOTECHNOLOGY: AN INTERNATIONAL ANALYSIS (U.S. Congress, Office of Technology Assessment 1984), at 309.

4. A. ANTEBI and D. FISHLOCK, BIOTECHNOLOGY: STRATEGIES FOR LIFE (1985), at 187. The patent was subsequently subject to some controversy, but in 1984 another process patent was granted with a lifetime ending in 1997. Ibid.

5. Abelson, *World Competition in Biotechnology*, 240 SCI. 701, May 6, 1988 editorial.

6. Crawford, *Biotechnology Investment Lags*, 241 SCI. 284, Jul. 15, 1988.

7. Antebi and Fishlock, *supra* note 4, at 183–221 *passim*.

8. GENE-SPLICING WARS, *supra* note 1, at 232.

9. *See* B. ZIMMERMAN, BIOFUTURE: CONFRONTING THE GENETIC ERA (1984), at 71.

10. Zimmerman, *Epilogue: Looking Ahead*, in GENE-SPLICING WARS, *supra* note 1, 241, at 242. Burke Zimmerman, Ph.D., was teaching Medical Ethics in a Public Policy setting in California at the time of writing.

11. Brill, 236 SCI. 1159, Jun. 5, 1987 letter.

12. COMMERCIAL BIOTECHNOLOGY, *supra* note 3, at 489–497 *passim*.

13. Krimsky, *Social Responsibility in an Age of Synthetic Biology: Beyond Biohazards*, in GENE-SPLICING WARS, *supra* note 1, 225. *Also see* J. CHERFAS, MAN-MADE LIFE: AN OVERVIEW OF THE SCIENCE, TECHNOLOGY AND COMMERCE OF GENETIC ENGINEERING (1982), at 134–139. For commentary reflecting attitudes in the late eighties, *see* Koshland, *Frontiers in Recombinant DNA*, 236 SCI. 1157, Jun. 6, 1987 editorial. For a detailed history of the technological issues, *see* EVOLUTION OF TECHNOLOGICAL ISSUES, *supra* note 2.

14. *See*, e.g., Fredrickson, *The Recombinant DNA Controversy: The NIH Viewpoint*, in GENE-SPLICING WARS, *supra* note 1, at 13.

15. 41 FR 27902, July 7, 1976.

16. Ibid., at 27902 and *passim*.

17. Nader, *Technology and Democratic Control: The Case of Recombinant DNA*, in GENE-SPLICING WARS, *supra* note 1, 139, at 145 *ff*. *See also* Zimmerman, *Science and Politics: DNA Comes to Washington*, in GENE-SPLICING WARS, ibid., 33, *passim*. A public forum was held at NAS in March of the same year. That forum drew not only the news media and protests from special interest groups but also scientific antagonists. Zimmerman, at 38.

Clair Nader, Ph.D., wrote as a science policy consultant in Washington, D.C.

18. 42 FR 49596, September 27, 1977. For a succinct history of events, *see* Karny, *Regulation of the Environmental Application of Biotechnology*, Spring 1987 ENVT'L LAW 1. Mr. Karny, a member of a Washington, D.C., law firm, identified himself as a specialist in biotechnology law.

19. Fredrickson, *supra* note 14, at 16.

20. Ibid.

21. Nader, *supra* note 17, at 151.

22. Krimsky, *supra* note 13, at 234.

23. Zimmerman (1984), *supra* note 9, at 8.

24. Ibid., at 164–165.

25. Ibid., at 165. In the end, Zimmerman reports that neither local ordinances nor Maryland's statute was being enforced in the early eighties. Ibid., at 176.

26. GENETIC ENGINEERING, *supra* note 2.

27. Ibid., at 19. Appendix 4 of this report provides a summary of Asilomar, which had previously been published through NAS. Appendix 10 provides a draft Environmental Impact Statement prepared by NIH pursuant to the National Environmental Policy Act (NEPA), 42 USCA 4321 *et seq.* The cases against genetic engineering and against allowing fear to deprive us of rDNA benefits are presented in separate appendices. *See* Note 2.

28. Fredrickson, *supra* note 14, at 40.

29. Zimmerman (1984), *supra* note 9, at 48–51.

30. Ibid., at 169–175, *passim.*

31. Ibid., at 172–174.

32. COMMERCIAL BIOTECHNOLOGY, *supra* note 3, at 355.

33. Ibid.

34. Zimmerman (1986), *supra* note 10, at 41–43 *passim.*

35. Zimmerman (1984), *supra* note 9, at 147.

36. Krimsky, *supra* note 13, at 225.

37. For an excellent selection of those statutes, *see* SELECTED ENVIRONMENTAL LAW STATUTES, 1990–91 Educational Edition (West Publishing Co.). For a synopsis of the contents of relevant statutes, *see* M. WOROBEC and G. ORDWAY, TOXIC SUBSTANCES CONTROL GUIDE (1989). Note that much of rDNA technology is regulated as chemical technology, with a focus on the biological or biochemical products of rDNA methods.

38. *See* note 27, *supra.* For an analysis of the role of NEPA's Environmental Impact Statement in the advancement of innovative technologies such as rDNA, *see* Binder, *NEPA, NIMBYs and New Technology*, 25 L & W L. Rev. (1) 11 (1990). Genetic engineering is specifically addressed at pages 30–33. The citations list cases in point and legal articles addressing rDNA, specifically. For then-current federal policy and regulation, *see also* BIOTECHNOLOGY: NEW DEVELOPMENTS IN FEDERAL POLICY AND REGULATIONS (Practicing Law Institute 1988) and citations.

39. 42 U.S.C.A. 4332 (2)(C).

40. Karny, *supra* note 18.

41. 51 FR 23302, June 26, 1986.

42. *Coordinated Framework for Regulation of Biotechnology: Hearing before the Subcomm. on Investigations and Oversight and the Subcomm. on Natural Resources, Agricultural Research and Environment and the Subcomm. on Science, Research and Technology of the House*

Comm. on Science and Technology, 99th Cong., 2d Sess. (1986), and *Issues in the Federal Regulation of Biotechnology from Research to Release*, SUBCOMMITTEE ON INVESTIGATIONS AND OVERSIGHT, HOUSE COMM. ON SCIENCE AND TECHNOLOGY, 2D SESS. (COMM. PRINT 1986). In 1988 the Office of Technology Assessment released a report on field testing rDNA organisms. Although it was received in the scientific community as an improvement on an earlier draft, the document was criticized as a scientifically naive set of compromises. *See* Davis, *Paranoia Invades Report on Man-Made Microbes*, WALL ST. J., Jun. 6, 1988, at 22.

43. *See*, e.g., Karny, *supra* note 18, at 2.

44. *See* 51 FR 23336–23393.

45. *See* 51 FR 23319–23335.

46. 7 U.S.C.A. 136–136y.

47. 15 U.S.C.A. 2601–2671.

48. Karny, *supra* note 18, at 2.

49. Ibid., at 2–3.

50. 15 U.S.C.A. 2604.

51. 4 INSIDE EPA'S POL'Y ALERT (5) 1, Mar. 11, 1987.

52. WALL ST. J., Oct. 21, 1987, at 1. For a later commentary on the intersection where rDNA meets the political system, *see* Melloan, *Learning Political Science at Crop Genetics*, WALL ST. J., May 10, 1988, at 39. The chairman of Crop Genetics International was John Henry, a Harvard graduate and former member of a senatorial staff.

53. BIOTECHNOLOGY:DEVELOPMENTS, *supra* note 38, at 105.

54. Fredrickson, *supra* note 14, at 19.

55. Zimmerman (1984), *supra* note 9, at 106.

56. *See* Karny, *supra* note 18, at 2.

57. Ibid. These decisions are Foundation on Economic Trends v. Heckler, 587 F. Supp. 753 (D.D.C. 1984), and 756 F2d 143 (D.C. Cir. 1985). They are also reported in Zimmerman (1986), *supra* note 9, at 48.

58. Foundation on Economic Trends v. Johnson, 25 ERC 1429 (D.D.C. 1986).

59. Foundation on Economic Trends v. Thomas, 25 ERC 1431 (D.D.C. 1986). Jeremy Rifkin was still embroiled in his efforts in 1988. His Foundation on Economic Trends had brought suit against NIH to force the preparation of an Environmental Impact Statement for rDNA research on AIDS. Booth, *Of Mice, Oncogenes, and Rifkin*, 239 SCI. 341, Jan. 22, 1988 News & Comment.

60. *See* Binder, *supra* note 38, at 41.

61. Singer, *Recombinant DNA Revisited*, 209 SCI. 1317, Sep. 19, 1980 editorial.

62. Karny, *supra* note 18, at 4.

63. *Science Wins One*, WALL ST. J., Jan. 14, 1988 editorial, at 26. *See also* Sun, *Preparing the Ground for Biotech Tests*, 242 SCI. 503, Oct. 28, 1988 News & Comment, describing relevant public relations efforts. *But see* Abelson, *World Competition in Biotechnology*, 240 SCI. 701, May 6, 1988 editorial, predicting the United States will be a second-rate player in international agricultural rDNA competition by reason of regulatory inhibitions. *See also* Sun, *Designing Food by Engineering Animals*, 240 SCI. 136, Apr. 8, 1988 News & Comment. Representatives of NAS were criticizing the administrative agencies for stifling innovation in the food industry.

64. Marshall, *Biotech Companies Lobby for Federal Regulation*, 248 SCI. 546, May 4, 1990 News & Comment.

65. Krimsky, *supra* note 13, at 231.
66. Zimmerman (1986), *supra* note 10, at 36.
67. Halvorson, *The Impact of the Recombinant DNA Controversy on a Professional Scientific Society*, in GENE-SPLICING WARS, *supra*, note 8, 73, at 73 and 87. Harlyn O. Halvorson was writing as the Director of Brandeis University's Rosentiel Basic Medical Studies Research Center.
68. Nader, *supra* note 17, *passim*.
69. COMMERCIAL BIOTECHNOLOGY, *supra* note 3, at 492.
70. Halvorson, *supra* note 67, at 75.
71. Ibid., at 78.
72. Michael, *Who Decides Who Decides? Some Dilemmas and Other Hopes*, in GENE-SPLICING WARS, *supra* note 1, 121, at 125–126.
73. Schmeck, *Recombinant DNA Controversy: The Right to Know—and to Worry*, in GENE-SPLICING WARS, *supra* note 1, at 93.
74. Ibid., *passim*.
75. Krimsky, *supra* note 13, at 231.
76. Zimmerman (1986), *supra* note 10, at 42–43.
77. Michael, *supra* note 72, *passim*. *See also* Halvorson, *supra* note 67, at 85.
78. Michael, ibid., at 124–125.
79. Zimmerman (1984), *supra* note 9, at 163.
80. Schmeck, *supra* note 73, 105.
81. Zinder, *A Personal View of the Media's Role in the Recombinant DNA War*, in GENE-SPLICING WARS, *supra* note 8, 109, at 117.
82. Halvorson, *supra* note 65, at 87.
83. Fredrickson, *supra* note 14, *passim*.
84. Thornton, *Commentary*, in GENE-SPLICING WARS, 29, at 30. Ray Thornton, president of the University of Arkansas in 1986, had been chairman of NIH's rDNA Advisory Committee in 1979.
85. Zimmerman (1984), *supra* note 9, at 247.

PART III

AT THE INTERSECTION: DISENFRANCHISEMENT AND PROSPECTS FOR REENFRANCHISEMENT

CHAPTER 9

At the Intersection: Merger or Collision?

THE CULTURAL CONTEXT

> "Without law there are no enforceable standards dealing with any scientific problem; without science there are no feasible means to resolve conflicts that inevitably accompany the many technical aspects of civilization."[1]

It might once have seemed likely that improved access to science and technology would have narrowed the issues set before those responsible for policy formulation in this country. Instead we have entered an era when issues have proved the most divisive and intractable in our history. We can no longer assume that our institutions will allow full airing of controversies with a consequent emergence of a public policy serving both individual and societal interests.[2]

In the preceding chapters we have considered elements of public participation in policy formulation in order to elucidate the actual results of blending science and technology into our politicolegal institutions. Our mission has been explanation as a foundation for predicting consequences whenever a new intersection looms. Predictions, when accurate, lead to proposals for improvements in the management of future integration among relevant disciplines.

Along the way, we have encountered substantial impediments to uneventful yet productive accommodations. On the one side is the impression that our political system is being directed by special interests in collaboration with a subverted press. On the other side is the human propensity to fill in gaps in science understanding with imaginative speculation, which displaces and more accurate comprehension of principles. Moreover, the two systems are burdened by a disturbing extension of the gap in communication between the humanities and the sciences. The lack of communication among interested professionals returns to society at large in the form of major lapses in communication with the lay public.

Many questions flow from these impediments to sound policy formulation in association with technological advances. The essential question is whether our educational, commercial, professional, and political institutions can function for the cooperative advancement of science and policy inextricably mingled. In short, do our traditions work at the intersection, or has the public in effect been disenfranchised?

Implications are that the public has been disenfranchised,[3] not by some diabolical coup d'état but by an intricate mix of inertia, paranoia, and widespread egocentrism, the last exemplified by inconsistent demands regarding pharmaceutical innovations on the individual level and the NIMBY ("not in *my* backyard") syndrome in environmental circles. All contributions to policy formulation are subject to distortion as they echo and re-echo through our institutions. Cynicism is the end product, which can serve only to further retard sound resolutions.

The essential question before us can be broken down into smaller, but no less difficult, parts: Who decides who is to make decisions of policy and law? How is ignorance to be alleviated? What are the legitimate interrelationships among professionals in the affected disciplines? How are fundamental unknowns or internal substantive controversies to be resolved in due course while being managed in the here and now? Can the requisite fine distinctions be effected through the combined forces of the marketplace and the politicolegal system?

An overarching philosophical mission must be the elimination of mysticism attributed to technology and, especially, science. The basic humanity of scientists and of the scientific endeavor is a basis for building the necessary bridges.

Whenever the marketplace either presents a problem or fails to offer solutions, we turn to government. We have seen that our politicolegal institutions can facilitate or impede technological advances through a variety of mechanisms. Presumably, the resultant decisions serve both individuals and society at large. It is also presumed that the public is the ultimate instrument of policy formulation. But is that the case at the intersection?[4]

In this context the needs attributed to the electorate include discovery of new resources, invention or substitution for resources in short supply, reduced costs of production, and the design of new products along with a corresponding creation of needed jobs. Perversely, deterrents to the satisfaction of perceived needs include the negative results associated with the positive steps taken toward the very goals identified.[5]

NATIONAL POLITICAL PHILOSOPHY AT THE INTERSECTION

> "There is virtually no aspect of government that does not involve science and technology."[6]
>
> "The best way the science advisor can serve . . . is to provide the most objective analysis of the scientific evidence surrounding important government decisions and to present these facts . . . without political considerations."[7]

Social regulation, as described in chapter 1, is one means for alleviating the negative influences of technological innovation in society, but regulation of or through technology is subject in its turn to substantial criticism. On the one hand, critics of social regulation at the intersection observe that the programs often fail to provide the minimum protection needed, while on the other, complaints of overregulation are frequent. Congress, it is asserted, has created bad legislation, and the courts are called upon to do what they neither can nor should do. Agencies suffer from conflicts of interest. Regulators often conduct inadequate investigation, then stall in going forward, to play it safe. Value judgments are confused with economic or scientific considerations. The overall result is a random array of decisions, with the balanced decision a rarity.[8]

Controversy in this arena, then, is hardly novel and can come as no surprise, especially when so many technologies having social consequences are instituted with little public discussion. That tax dollars may have supported the process seems to make little difference.[9] It is inevitable that controversy should follow and is actually the more vehement because public entry not only has been put off but is also perceived as coming too late for accommodation.

Most notable among these controversies are those surrounding nuclear power technology and the arms race. Other technological advancements have known their share of controversy among the public, as has been illustrated at length throughout this book.

Surprisingly enough, in 1979 there was an international consensus to the effect that ordinary citizens do appreciate science and technology to a far greater extent than most would credit. Nevertheless, the companion consensus was that anxiety results from a sense of powerlessness, expressed as a call for more direct influence on policy formulation.[10] That impotence, largely self-imposed, was attributed to perceptions of a closed government in combination with inadequate or inaccurate "education" on the personal level and at the hands of the variety of educational processes themselves.[11]

During this country's 1988 presidential campaign, the significance of the intersection was expressly recognized. The Republican platform held that national progress depends on scientific and technological innovation, an "economic fountain of youth." Prerequisite to progress was excellence in education, with special reference to scientific illiteracy.[12] It was the first time science and technology were afforded more than a minor place in the thinking of our political parties. One basis for the increased visibility was a briefing of the Republican platform committee by three prominent scientists. The significance of the event is only slightly reduced by the fact that all three represented only one science, physics.[13]

Less than a year before, a voice from academia had asserted that our social structure had changed just since the sixties to one based in knowledge and education rather than in ownership of land and machinery or in institutional ties such as labor unions. A contemporaneous survey, however, found that more than half the population believed that public officials cared little about what any individual thought. The number holding that unfortunate belief had more than doubled between 1960 and 1986.[14]

As has been explored in some detail in the preceding chapters, the major political intersections are found where codified law is being drafted, implemented, and enforced. Today:

> It is mandatory that every young scientist, and indeed every educated person, [become acquainted] with at least the over-all environmental processes and conditions that make possible the very survival . . . of individual organisms . . . In a democracy it is not sufficient just to have a few trained persons who understand . . . ; there must also be an alert citizenry.[15]

Whatever position one takes on any given technological matter, legitimate questions *are* raised on behalf of the electorate. These may not be, however, the questions raised by the several partisans. For the two to merge, individuals must prepare themselves through greater awareness.

The questions that are—or should be—raised on behalf of society deserve to be clearly set out:[16] How are we to deal with any new knowledge with profound implications? What is it that determines perceptions of risk or of safety? Why do those perceptions change, even in the absence of data allowing scientific resolution?

Answers to all of these questions are crucial to maintaining a democratic republic in face of issues constantly arising from sophisticated innovations in science and technology. Unfortunately, no answer is readily discerned. Further, many would pessimistically respond in the negative to the overarching question regarding the capacity of our institutions. Few of the observers propose solutions with any practical detail, let alone practicability. Seldom is the existence of a variety of public voices taken into account.

None of us should be prepared to accept any situation so untenable. Every member of the electorate bears some responsibility in finding answers at one level or another. There is, as Eugene P. Odum suggested almost three decades ago, one generalized answer to the questions surrounding effective disenfranchisement: *knowledge.*

Knowledge in this country is instilled through two primary tools, formal education and the press. These tools, however, if not perverted, are clearly not accomplishing their implicit mandate, at least in the arena of concern here. Most members of the present electorate, having completed or merely left the system of formal education, depend heavily on the news media, apparently to little avail, despite the need to be empowered by knowledge if they mean to govern themselves. In the case of science, public perception comes almost exclusively through the press, which, significantly, can choose between responsibility and the sensational.[17] It is the sensational that "sells."

Paramount among the questions at hand is who is to decide who will make the decisions. But a prerequisite to pursuing answers to this and all the related questions, express or implicit, is acknowledgment of the tension in this country between the public and private sectors. It has existed for no less than fifty years, but the proper size and role of each sector continually undergoes reevaluation.[18]

Another preliminary consideration is reiteration of the demand of democratic processes for wide-ranging public discussion to guide strategies and operations. This demand requires someone to represent the public interest in the first instance, and it requires the establishment of priorities wherever multiple specialized interests are involved. Here it must be kept in mind that, for better or worse, the influence of self-initiated citizen advocates (primarily the "public interest" groups) does not necessarily arise from political or economic power or from a position of superior knowledge. Legally, that influence arises exclusively from rights guaranteed by the First Amendment. Even in face of public pressures, it is essential that the positions being proferred ultimately be accepted or rejected solely on the basis of their merits (including values, morality, and ethics). To choose solely on the basis of the identity of the active proponents would be a grave mistake.[19]

LAWMAKING AT THE INTERSECTION

> "[T]here is not a single 'public' but many publics, and . . . rarely on even the simplest issue can one truly speak of *majority* wishes or interests."[20]

A great many difficulties arise when policy decisions rest in formal efforts to grapple with science and technology. The first manifestation of these decisions usually takes the form of legislation, followed in succession by regulation, enforcement, and adjudication.

Although there is a certain consistency in the pattern of evolving policy, the successive steps do not always follow in this orderly fashion. Whatever the progression or degree of order, however, the underlying process by which facts are produced and considered must at the very least *appear* to be fair. In the world of politics and policy, scientific "information can easily become a weapon deployed by adversary-parties to formal actions. Moreover, [r]egulation is unsatisfactory partly because we have not faced up to the science and technology part of the task."[21]

The various officials involved in policy formulation obviously know that decisions are to be based in fact, but under the pressure to take some form of "timely" action, they must often act on the basis of "best available" information, which can be uncertain, incomplete, or both. Decisions, then, are based not only on whatever factual bases may exist but also on a "large variety of political and policy considerations that often go unvoiced."[22]

More than one paradox can follow. The federal pollution control program provides numerous instructive cases in point. Two examples are most revealing, both of the regulatory arena. The first paradox is that regulatory action can appear to have exactly the opposite from the intended result. A good example is the offset requirement in federal air pollution control. The mandate calls for the reduction of existing pollution before superior facilities can be constructed in a given area. For practical reasons it may not be possible to reduce pollutants from existing plants. As a result, the mandate in practice discriminates against updated, efficient, cleaner facilities. Heavily polluting old facilities remain, because new plants will not be established. Further, investment in new technologies for cleaner plants is discouraged.[23]

The second example is comprised of the negating activities and indecisiveness of agencies in dealing with technological activities. When critics of responsible agencies are politically powerful, that criticism can ultimately cause agencies to depart from their protective missions in order to appear more evenhanded. In so responding, missions are compromised.[24]

All of these circumstances will be exacerbated in those cases where corruption or favoritism is a reality. Although the laissez-faire of the nineteenth and early twentieth centuries remains out of fashion, it has been argued that at least some political abuses could be eliminated through greater reliance on the competitive market, at least in times of prosperity.[25]

We have already seen that Congress is ill-suited to address issues of science or technology in a nonpolitical manner but that delegation fails to resolve the quandaries besetting our elected representatives. Delegation merely shifts unresolved matters for the decision within the agencies,[26] in actuality seldom any resolution at all.

THE JUDICIARY AS A REFLECTION OF PUBLIC VOICES

> "Whether courts have . . . *creatively* responded to science-policy disputes, whether and to what extent they may *legitimately* do so, and whether

courts are *institutionally equipped* and *capable* of auspiciously responding, remains problematic."[27]

Thus far, the focus of discussion has been directed primarily to policy formulation through federal legislation and regulatory processes. The judicial system has, for the most part, been mentioned only in passing. But it is here that the voices of society may be most nearly ubiquitous and have the effects most immediately felt. It is here that conflicts associated with legislation and regulatory processes become obvious. Those which have remained submerged in the course of events are finally brought to the surface.

In many ways the judiciary, clearly isolated by constitutional decree and reality from its governmental partners, is itself outside science and technology as well. The legal system may be the branch of government most overtly moved by the quandaries met at the intersection. Before turning to the courts, it is worthwhile to reset the administrative stage our judiciary is reviewing. Programs administered by EPA are typical.

In the decade since the National Environmental Policy Act (NEPA)[28] entered the national consciousness, some 30,000 proposed federal actions were reviewed annually for a determination of environmental consequences potentially demanding at least an Environmental Impact Statement. Some 8,000 Environmental Impact Statements had been filed. Here was one of the first opportunities for significant public participation in an interdisciplinary arena, with its point of departure founded in environmental sciences. But the sad conclusion is that the associated public hearings have lost any meaning in terms of relevance and public comprehension of the matters at hand. Many explanations have been offered, but even the casual observer would take immediate cognizance of a truly incredible volume and complexity in the administrative records.[29] The NEPA experience is one more example, along with pharmaceutical innovation and the more specific issues of recombinant DNA, of the heavy technological burden taking codified policy formulation into our courts.

Difficulties experienced through the complexity and sheer volume of relevant information can only be exacerbated when the adversarial process encourages the advocates, in keeping with their background and training, to seek more to conquer the truth than to discover it[30]—a task most attorneys and all litigators embrace with considerable relish.

Risk assessment and other intellectual technologies[31] may have drawn attention first in pharmaceutical innovation and be most sensationally a part of the recent uproar over molecular biotechnologies, but these technical tools are entrenched throughout environmental protection movement. The result may be best expressed in the report and supporting appendices published in 1990 by EPA's Science Advisory Board (SAB).[32]

For all the notable successes over the first two decades of the evolving federal program, the SAB concluded that hindsight actually projected an image of inconsistency and lack of coordination, resulting in a fragmentary policy.[33]

A synopsis of findings should sound disturbingly familiar. EPA program offices, responsible for implementing specific statutes

have tended to view environmental problems separately; each program . . . has been concerned primarily with those problems that it has been mandated to remediate, and questions of relative seriousness or urgency generally have remained unasked. Consequently, at EPA there has been little correlation between the relative resources dedicated to different environmental problems and the relative risks posed by those problems.[34]

On the basis of its review of EPA programs, the SAB made ten recommendations for improvement. Notable among them are returning to the protection of natural ecosystems, continual improvement of underlying scientific data and analytical methodologies, explicit comparison among environmental risks, the setting of priorities among risks, improved environmental literacy, and integration of environmental approaches.[35]

More specifically, the SAB observed, in the context of efforts to rank risks, the absence of data in some instances and substantial uncertainty in the data elsewhere. The SAB concluded, "[w]ithout more and better data, conclusions about relative risk will be tenuous and will depend in large measure on professional judgment."[36] Further, the report challenges present political response to public voices, as we endeavor to merge environmental with economic principles:

Reliance on "willingness to pay" and similar techniques commonly used in economic analyses has distorted current understanding of natural resources. While some people may not care about wetlands and assign no value to their existence, such areas still provide valuable ecosystem services to this and future generations. While few people are likely to care about and be willing to pay for plankton and fungi, such organisms play a critical role in sustaining economically valuable ecosystems.[37]

Appendix A of the report made the same statement even more forcefully:

When it is applied to the valuation of ecosystem services, the assumption [of economic analysis] that value derives from individual preferences may be inconsistent with fundamental ecological principles. Individuals may enjoy the benefits of these services without any knowledge of their existence, thus their preferences may imply values that do not reflect the ecological importance of natural systems and the services they provide to humans.[38]

Any resolution of this sort of quandary is beset by the inevitable—and entirely appropriate (so long as identified as such and based in accurate information)—value judgments. It is here and under these circumstances that the citizen should possess the most persuasive voice.

It may be that public participation at the intersection should delegate the technological and interdisciplinary realities to the professionals and the niceties of law to their governmental representatives, subject only to public values, constitutionally prescribed, until such time as scientific illiteracy can be effectively mitigated, if not eliminated altogether. As the SAB discovered, probably to no surprise among practicing environmental professionals, "environmental risks considered most serious by the general public today are different from those considered most serious by the technical professionals charged with reducing environmental risk."[39] We are, then, brought back to whether public policy should be responsive to public needs or to public demands.

This time, those raising the issue provide the opening for a meaningful answer in a forum that may actually be heeded: "One way to bridge this dichotomy would be to improve the public's understanding of the scientific and technical aspects of environmental risk while improving scientists' understanding of the basis of public concern."[40]

The common sense behind this answer is most appealing. Here is a call, official, informed, and even representative, for both education and integration to effect policy formulation that is responsive to public needs without abrogation of those pesky interdisciplinary realities.

In this country and at this particular juncture, it may well be the dynamic elaboration of environmental law and policy that most completely exemplifies the situation at the intersection. And this dynamic is most obviously manifested in judicial review of policy evolving through law. Once more the quandaries summarized at the beginning of the chapter are manifested.

There can be no doubt of the importance of the judiciary in policy formulation at the intersection. Complicated scientific and technological disputes have been in court throughout Anglo-American legal history. Prominent regulatory roles, however, have emerged since the seventies, and the judicial system has been increasingly involved in a supervisory capacity[41]—at least after the fact of statute, regulation, implementation, or enforcement.

As is the emerging norm, the associated quandaries at the intersection are based both in the judicial institutions themselves and in the reception and application of science and technology through triers of fact and of law. By their very nature, judicial decisions are necessarily limited, fragmented, and incremental,[42] on a parallel course, in fact, with scientific process and, not insignificantly, with the codifications under review. Notably, in mathematics as elsewhere, parallel courses are destined never to meet, let alone merge. Significantly enough, the courts have themselves been charged with a mysticism comparable to that attributed to science.[43] Here again, another charge arises to the effect that the legal system discourages scientific and technological innovation, this time illustrated through emergent trends in tort causes of action.[44]

Going beyond the internal affairs of the court system as such, the fact is undeniable: The judiciary was not designed for and has not proven amenable to resolution of scientific disputes arising within its own forums or transferred from administrative processes.[45] The incapacity is felt in a number of ways affecting all interested sectors.

At the outset, judges (and reviewing justices) are not endowed with inherent scientific knowledge. Neither are they ordinarily trained in scientific thinking or processes. They, in short, have no *scientific* basis for establishing the reliability of putative scientific or technological testimony, evidence or, ultimately, remedies.[46] Sometimes pure logic, however skillfully applied, is not enough. Triers of fact must be prepared to probe the validity of the reasoning behind expert conclusions and to hold those expert witnesses to standards and criteria prevalent in their respective fields of endeavor. Furthermore, the limits of each discipline must be accepted in a meaningful way. None of these dictates, according to attorney Bert Black, is being heeded in our courts.[47] As a result, the "inexpert" expert holds sway.[48]

Despite widespread acknowledgment of problems at the intersection, especially in the courts, little institutional or procedural reform was perceived as recently as 1986.[49] It is unlikely that any is forthcoming in the immediate future. As Shiela Jasanoff observed,

> Judges have become almost as powerful as legislators in shaping science and technology policy, but they approach the task with intellectual and material resources that are hardly different from those available . . . fifty years ago. And although lawyers and scientists frequently face each other in judicial . . . proceedings, they remain, for the most part, uninformed about each other's professional goals, assumptions, and patterns of reasoning.[50]

One result, profound in its implications, is the admission of scientific testimony that is more misleading than enlightening and may be worthless. Such testimony is admitted in some instances simply because scientific errors, not apparent to nonscientists, escape any efforts by opposing counsel to enlighten the court. Judicial reasoning may be further clouded by popular misconceptions added to the complexity and even novelty of the evidence itself.[51] These problems persist despite the fact that scientists, lawyers, and the public are all motivated by the common mission of resolutions that are equitable, efficient, and *competent.*[52]

On the practical level and with grave implications for future exchanges, the experts themselves come away from the courtroom with the sense of having been manipulated by the advocates and by the institution itself. In part, this reaction is attributed to the superimposition of advocacy on the presentation of information.[53]

Presumably, this imposition is layered over any scientific controversy as such. No one who has ever read the letters in *Science* or attended a conference of scientists can doubt that scientists themselves advocate their *scientific* positions in the strongest terms possible and are hardly above the occasional ad hominem attack. Without controversy at this level, science would not advance. The problems arise when a nonscientist presumes to direct the controversy from a position external to scientific methods or principles. A systemic problem for the intersection is that, while scientists can work forever with issues remaining unresolved, a court—and the politicolegal system, if not the marketplace—must act in the present and preferably in black or white, without the greying effect of the qualifications scientists demand.[54]

Despite some very real problems, there are those who would demand a shortcut to resolution that essentially denies the restrictions current science and technology impose. In recent years the shortcut has taken the form of legislative proposals for victim compensation. Triggered in large part by Section 301(e) of CERCLA,[55] the approach is to substitute a legal presumption, rebuttable or absolute, for medical causation in the individual and for traditional legal causation linking plaintiff to the harm and its source, the "tortfeasor."[56]

As the sponsor of one such "toxic victims compensation act" insisted:

> In toxic tort actions the obstacles [to recovery by plaintiffs] are not only legal; they are founded in the failings of modern science. The state of scientific knowledge and the art of medical diagnosis do not always permit legal or scientific conclusions about the causes of a particular disease. Indeed, the rigors of the scientific method itself requiring repetition and peer review, may prevent the timely generation of proof needed to support the evolution of common law remedies.
>
> Simply put, the law has not yet developed adequate mechanisms to protect the toxic victim.[57]

It requires the incisive mental processes of neither a scientist nor a lawyer to recognize that such a "resolution" to the dilemmas at the crossroads serve none of the participants, including society at large. Law and, certainly, science have not failed at all. But what science today cannot do for law is distinguish *on demand* who is a "toxic victim."

The list of inequities likely to follow the simple expedient of ignoring the scientific realities finds victims—and windfalls—on all sides. As the mixture of potential causes, including the rich mix of natural and synthetic chemicals to which each individual is exposed in a lifetime, is scrutinized in greater detail, the more difficult the task of isolating sources of morbidity and mortality becomes. Bludgeons are not the answer. Knowledge and respect for the principles of all the institutions are.

After all, as a former member of EPA's Office of General Counsel observed,

> Everyone dies eventually and generally suffers ill health before they do. Nature, beyond its own innate cycles, is under constant pressure from development, human population, and overharvesting, pollution damage quite apart. Control devices and

techniques very generally do not work as well as they should, or fail in unexpected and self-defeating ways. Such a perception may seem commonplace. But, in fact, it often is ignored in environmental debates. All sides seem to share a common interest in making the problems appear either not too significant, or soluble by the controls at issue.[58]

The lesson which must be taken to heart is that, while law can be and is accommodated in scientific and technological circles, we and our politicolegal system must learn to accommodate science and technology. Even when the realities are frustrating, we must work with, not against, the scientific revolution on behalf of individuals and the larger society. Until we do, the electorate will remain disenfranchised, overtly and with more insidious subtlety. Our policy formulation will linger on in a state of paralysis.[59] We are almost certainly unable to come up with a better system; what is needed are approaches at all levels that are scientifically credible, legally defensible, and politically feasible.[60] Recent history, especially with the Food and Drug Administration and the EPA, tells us we have set no easy course for ourselves.

NOTES

1. Thornton, *Uses and Abuses of Forensic Science*, in SCIENCE AND LAW: AN ESSENTIAL ALLIANCE (W. Thomas, ed., 1983), at 79. John A. Thornton was a professor of forensic science at the University of California, Berkeley.
2. In addition to the foregoing materials, *see* T. GREENWOOD, KNOWLEDGE AND DISCRETION IN GOVERNMENT REGULATION (1984). Ted Greenwood was in the Office of Technology and Policy during the Carter administration.
3. *See*, e.g., Nader, *Technology of Democratic Control: The Case of Recombinant DNA*, in THE GENE-SPLICING WARS (R. Zilinskas and B. Zimmerman, eds., 1986), 139.
4. For a reiteration of the issues as articulated on an international scale at the close of the seventies, *see* K. NICHOLS, TECHNOLOGY ON TRIAL: PUBLIC PARTICIPATION IN DECISION-MAKING RELATED TO SCIENCE AND TECHNOLOGY (Organization for Economic Cooperation and Development 1979). Among the twenty-three members of this international organization were Australia, Belgium, Finland, France, Italy, Japan, Portugal, and England, as well as the United States.
5. Ramo, *Regulation of Technological Activities: A New Approach*, in SCIENCE AND LAW, *supra* note 1, at 35. Simon Ramo is the founder of TRW, Inc., and a member of the National Academy of Sciences.
6. Presidential candidate George Bush, *Science Policy*, 242 SCI.173, Oct. 14, 1988 Policy Forum.
7. Presidential candidate Michael Dukakis, Policy Forum, ibid.
8. Nader, *supra* note 3 *supra*.
9. Ibid., at 140.
10. Nichols, *supra* note 4, at 11–12.

11. Ibid., *passim.*

12. Bush, *supra* note 6, at 174.

13. Byrne, *Science and the Party Platforms,* 241 SCI. 1152, Sep. 2, 1988 News & Comment.

14. Rogers, *Similarities Exist in 1988 and 1960 Elections, But Big Contrasts Lie in Voters, Political Process,* WALL ST. J., Dec. 27, 1987, at 46.

15. E. ODUM, ECOLOGY (1963), at 1. Eugene P. Odum is a pioneering ecologist, whose college text, *Fundamentals of Ecology,* first published in the late fifties, is still in print in its 1973 third edition. It is used in the Ecology and Environmental Sciences course at Vermont Law School's Environmental Law Center for J. D. candidates and master's students who have no background in the relevant sciences. It is telling that all of these students are well educated, some in a scientific discipline. Most enter the environmental law program without previous education in basic ecological principles.

16. These are the questions raised by the editors of GENE-SPLICING WARS, *supra* note 3, *passim.*

17. *Id.,* at xiv. For more on the role of the press and commentary on its modern manifestations, *see also,* e.g., Williams, *Legal Problems in Producing a Docudrama: Defamation and False Light Privacy,* 1984 ENTERTAINMENT, PUBLISHING AND THE ARTS HANDBOOK; B. BAGDIKIAN, THE MEDIA MONOPOLY (1983); D. HALBERSTAM, THE POWERS THAT BE (1979).

18. Grobstein, *Asilomar and the Formation of Public Policy,* in GENE-SPLICING WARS, *supra* note 3, at 3.

19. *See* Nader, *supra* note 3, at 143.

20. Wengert, *Environmental Policy and Political Decisions: The Reconciliation of Facts, Values and Interests,* in POLITICS AND ECOLOGY (P. Foss, ed., 1972), 40, at 45.

21. Ramo, *supra* note 5, at 38.

22. Levine, *Love Canal and the Limits of Scientific Proof,* 2 NAT. RESOURCES & ENV'T (2) 21 (1986). This issue of *Natural Resources & Environment* was dedicated to analyses of the "fallout" from "The Explosion of Science in the Courts."

23. Ramo, *supra* note 5, at 39.

24. Ibid., at 40–41. Simon Ramo cites as a prime example the Occupational Safety and Health Administration.

25. P. DOUGLAS, ETHICS IN GOVERNMENT (1952, 1972 reprint) at 31. When Senator Douglas was writing, the era of laissez-faire was still quite recent. Abuses attributed to "the massed power of big business" in combination with the "relative ignorance and weakness" of consumers, investors, and workers had led to enactment of the Pure Food and Drug Act and of many other social welfare statutes. Ibid.

26. In addition to the foregoing chapters and references, *see* Light, *Toxic Injury Compensation: One More Overview of Legislative Proposals,* 15 NAT. RESOURCES L. NEWSLETTER 2 (1983). Mr. Light was practicing in the Richmond, Virginia, law firm of Hunter & Williams and was secretary to the Ad Hoc Committee on Tort Law Reform of the American Bar Association's Corporate, Banking and Business Law Section.

27. O'Brien, *The Courts and Science-Policy Disputes: A Review and Commentary on the Role of the Judiciary in Regulatory Politics,* 4 J. ENERGY L. & POL'Y 81 (1983), at 81. (Emphasis in original). David M. O'Brien was an associate professor in the

Woodrow Wilson Department of Government and Foreign Affairs at the University of Virginia.

28. 42 U.S.C.A. 4321 *et seq.*

29. *See,* e.g., Nichols, *supra* note 4, at 81–92, *passim.*

30. Ibid., at 95.

31. *See* Brooks, *Intellectual Technology: The Dilemma of Environmental Law,* 15 RUTGERS COMPUTER 7 TECH. L.J. (2) 411 (1989).

32. REDUCING RISK: SETTING PRIORITIES FOR ENVIRONMENTAL PROTECTION (US Environmental Protection Agency, Science Advisory Board, SAB-EC–90–021 1990). The three appendices address, respectively, ecology and welfare (Appendix A), human health (Appendix B), and strategic options (Appendix C).

33. REDUCING RISK, ibid., at 1.

34. Ibid., at 3.

35. Ibid., at 6.

36. Ibid., at 8.

37. Ibid.

38. REDUCING RISK, APPENDIX A, *supra* note 32, at 31.

39. REDUCING RISK, *supra* note 32, at 12. Significantly, M.S.L. candidate Cynthia Argentine at the Environmental Law Center proffered an interesting approach in the course of Toxic and Hazardous Pollutants Law. Her suggestion that national referenda in point be conducted met with no oppositions among fellow J. D. and M. S. L. candidates. To avoid the excesses of the California model, however, this group seemed to concur that education through competing "fact sheets" should replace elaborate advertising campaigns.

40. Ibid.

41. O'Brien, *supra* note 27, at 112.

42. Ibid., at 113.

43. *See,* e.g., *Learning about Law on the Cheap,* WALL ST. J., Oct. 13, 1988, at Bl.

44. *See,* e.g., *Litigation Discouraging New Products,* 3 REFORMER (1) 1 (1988). *The Reformer* is published by the American Tort Reform Association.

45. O'Brien, *supra* note 27, at 114.

46. Black, *Evolving Legal Standards for the Admissibility of Scientific Evidence,* 239 SCI. 1508, Mar. 25, 1988. Significantly, Bert Black is an attorney at the Baltimore offices of Venable, Baetjer and Howard. The fact that he was writing for a scientific audience is an encouraging sign.

47. Ibid., *passim.*

48. *See,* e.g., Lettow, *Strategic Choices in the Selection of Expert Witnesses,* 2 NAT. RESOURCES 7 ENV'T (2) 6 (1986), and Gots, 6 REGULATORY TOXICOLOGY & PHARMACOLOGY 95 (Conference on Causation and Financial Compensation; Institute for Health Policy Analysis, Georgetown University Medical Center 1986). Charles F. Lettow was an attorney in the Washington, D.C., offices of Cleary, Gottlieb, Steen & Hamilton; Ronald E. Gots, M. D., J. D., heads the National Medical Advisory Service in Bethesda, Maryland. That these professionals are obviously communicating across professional lines is responsive to the first half of the larger problem before us.

49. Jasanoff, *Science and the Courts: Advice for a Troubled Marriage,* 2 NAT. RE-

SOURCES & ENV'T (2) 3 (1986). At the time of writing, Sheila Jasanoff was the director of Cornell's Environmental Law and Policy Program.

50. Ibid.

51. Gots, *supra* note 48, at 98.

52. Jasanoff, *supra* note 49.

53. Gillett, *Dilemmas of an Expert Witness*, 2 NAT. RESOURCES & ENV'T (2) 10 (1986), at 53.

54. *See* ibid., at 54.

55. Adequacy of Existing Common Law and Statutory Remedies, 42 U.S.C.A. 9601, *et seq.*, at 9651(e). CERCLA is the Comprehensive Environmental Response, Compensation, and Liability Act.

56. *See* Hassler, *Remedies for Victims of Toxic Torts: The Report of the Section 301(e) Study Group*, 15 NAT. RESOURCES L. NEWSL. (3) 1 (1983). *See also*, e.g., the April 13, 1983 "Dear Colleague" letter of Representative John J. LaFalce of New York. The congressman was seeking support of H. R. 2482.

57. Ibid., at 2. For another perspective, *see*, e.g., Ember, *Legal Remedies for Toxics Victims Begin Taking Shape*, 1983 CHEMICAL & ENGINEERING NEWS 11.

58. Pederson, *What Judges Should Know about Risk*, 2 NAT. RESOURCES & ENV'T (2) 35 (1986), at 37.

59. *See* Gaines, *Science, Politics and the Management of Toxic Risks through Law*, 30 JURIMETRICS J. (3) 271 (1990), at 273.

60. Ibid., at 314 and 275.

CHAPTER 10

Can We Reenfranchise Society for the Intersection?

THE CHALLENGES

> "It is of great importance that the general public be given the opportunity to experience . . . the efforts and results of scientific research. It is not sufficient that each result be taken up, elaborated, and applied by a few specialists."[1]

In this country the political process by which public policy is formulated can be represented in a series of deceptively simple steps. Ostensibly we proceed from information to political response to policy formulation. More realistically, policy *is* based on information, but that information is highly filtered. The first filter is that of the public's perception of the information. The second is politically motivated response to the perception. The third takes the form of governmental policy regulating a subject activity, more on the basis of the perception and political response to it than in keeping with the more distanced basic information. Furthermore, the whole process is purely ad hoc; that is, each regulatory quantum of filtered information is managed separately without connection to other quanta.[2]

In fact, we persistently restrict the relevant bodies of knowledge to small groups, each specialist in its own right, against the caveat of scientist-philosopher Albert Einstein. The result is as he predicted: a deadening of philosophical spirit and spiritual poverty,[3] or at least a deadening of the intellectual underpinnings of national spirit.

A modern scholar has observed that both (undergraduate) students and the nation would be well served through enough knowledge of science to provide the basis for future learning and the ability to listen intelligently to the arguments of experts. As F. H. Westheimer put it: "Lack of knowledge will not prevent [individuals] from having opinions; it will only prevent them from having informed opinions."[4] A more practicable observer called for four institutional responses: K. Guild Nichols's starting point was greater public accountability, but he would support accountability with freer public access to technical information. Access, in turn, would support more timely consultations on policy options. The fourth overall institutional response would be a more holistic approach to assessing those effects commonly characterized as "impacts,"[5] a term itself suggestive of crises resulting from scientific illiteracy in the face of the ongoing scientific revolution.

To the extent our institutions accept the reality of disenfranchisement at the intersection, two challenges must be taken up. In addition to the matter of scientific illiteracy, there is the challenge of distilling, from all the demanding voices purporting to represent the public, the actual societal needs (and then having the courage to act upon those needs[6]).

The Quandary of Scientific Illiteracy

> "The mistake . . . is the belief that we can technologize our way out of the crisis [of global environmental degradation], that is, that science and technology will rescue us from the consequences of stupidity, arrogance, and ecological malfeasance."[7]

There is a consensus in this country that we suffer not only from scientific illiteracy but also from a more fundamental general illiteracy. It is disheartening to be advised that college seniors are ignorant of philosophical, historical, and literary facts once taken as general knowledge.[8] Although not at issue here, general illiteracy is certainly a fundamental impediment to scientific and technological literacy. There is a vast literature at all levels on illiteracy, both general and in the sciences; here the focus remains on a small sampling of the latter.

Scientific illiteracy is of concern at two levels, both in preparing for a future heavily dependent on science and technology[9] and in assuring that the electorate is informed.[10] It is being asserted that we are failing at both levels. One definition of general scientific literacy is that all who finish high school be able to read newspaper and magazine items (such as those cited throughout this work) and be able to react to issues of science policy that touch their lives on a level more fruitful than simplistic emotional response.[11] The need is demonstrated most acutely where policy formulation requires choice between competing theories, action on the frontier of a scientific field and facing an array of experts with diverse positions.[12]

George W. Tressel, speaking as a representative of the National Science Foundation (NSF),[13] provided a succinct synopsis of the problem of scientific illiteracy in noting that, in the latter part of the eighties, we were teaching nothing about science in the primary grades. Then, at age 14, pupils were asked if they wanted to take on a course in basic science. Less than a quarter acquiesced, and most of them were taking into account college entrance requirements. No one was provided with an integrated overview fostering familiarity with science and technology at large. In short, neither the prospective technologist nor the general population was being served.

According to Tressel, then, what is needed is a comprehensive, integrated view of science introduced to students long before college age in order to produce not only technologists but also decisionmakers with a fluent perspective on science rather than exposure to an isolated course in one of the basic sciences.

Once the students in undergraduate studies enter scientific courses, they are met with programs asserted to be out-of-date, unimaginative, and inadequately organized to accommodate all the interests represented.[14]

Obviously, the problems are going to carry over into the areas highlighted here and be especially detrimental in the heavily scientific and technical fields of environmental protection and pollution control. One apparently simple choice, disposable versus cloth diapers, is representative. This deceptively simple problem lends itself to no simple answer. Instead, it actually "involves the great complexity of water consumption, detergent pollution, habitat destruction, pesticide use, energy consumption, exhaust emission, and waste disposal."[15]

This area of potential policy formulation is at least simple on its face. Most issue are far more complex even in their mere statement. One does not have to be illiterate in any sense to be justly accused of not recognizing individual contributions to major pollution problems. Only a few have any comprehension of the size and scope of the matters at issue in our environmental programs.[16] Without some foundation in the relevant disciplines the individual will be overwhelmed, unable to participate meaningfully in politico-legal forums, and in effect disenfranchised.

As a result of this effective disenfranchisement, the public may express concern about environmental matters in the broadest perspective, but individuals can focus their attention on only a few limited issues while matters having the broadest import go unnoticed.[17] As one educator chillingly observes, "The survival of the planet depends on whether future generations can be educated in ecological literacy—an awareness of the interconnectedness of all life. Such an education requires fundamental changes in many of our present assumptions about schooling."[18] This statement cannot be dismissed as hyperbole.

The Challenge of Distilling Public Needs from Public Demands

> "One enormous difficulty in [environmental] debate is the lack of a clear moral advantage . . . with either side. That deficiency has forced an un-

> pleasant need to examine the facts of the case, an activity for which Americans lack experience and find to be boring."[19]

With or without the fundamental problems posed by illiteracy, there is another major challenge for policy formulation: the need to extract, from the multitude of vociferous voices making inconsistent demands, the real public needs amenable to marketplace or politicolegal response. In recent years this requisite distinction has been acknowledged with reference to a silent majority, often out-shouted by vocal minorities as expressed by self-appointed advocates.[20] The concepts have become so pervasive that "Silent Majority" has attained the status of a proper name. Although it may be that demands and needs will never be coextensive, the dichotomy is exacerbated where, as with environmental matters, a major technological foundation exists.

Throughout the preceding discussions we have observed the multitudinous sources of public opinion, the most important being the lessons learned in formal educational settings as extended through both the news and the entertainment media. We have seen how individual interests can either be sacrificed to or assume a preeminence defeating societal interests. Yet Kim Herman Goslant has earnestly advocated direct, active citizen participation in environmental processes and has actually claimed success in the context of resource restoration. For this mission, however, he calls for a major educational component in the corresponding public interest program, and he distinguishes education from the more common component of propaganda.[21]

It is difficult to avoid concluding that much of what is happening in policy formulation is based on misguided voices arising at least in part from inaccurate perceptions of the matters at hand. That this charge has merit is largely confirmed in academic writings in point and in specific public movements to correct perceived inadequacies. One article by a sociologist serves to illustrate the former, while California's Proposition 65 is illustrative of the latter. The article reveals how the academician can actually contribute more to the problem than to its ultimate resolution.

Peter C. Yeager wrote of *perceived* biases in the enforcement of pollution control law.[22] In doing so he declined to offer pertinent definitions, such as "illegal business activity" and "corporate misbehavior," thus failing to recognize such fine distinctions as statistical excursions, inadvertent noncompliance, noncompliance based in unusual circumstances, knowing noncompliance, and outright violations, willful or otherwise.

As his thesis was based solely on statistical analyses with no apparent effort to uncover any factual matters behind those statistics, Yeager also omitted any reference to very real negotiation processes or to legitimate prosecutorial discretion, let alone any consideration of basic scientific, technological, or economic considerations, all of which go into enforcement decisions. Thus, he views the matter of bias as an isolated quantitative sociological phenomenon without reference to relevant disciplines among the natural sciences and technology or practicalities within politicolegal institutions.

Not surprisingly, Yeager discovered statistical evidence for bias on the part of this representative agency, but it was a perception this observer did not investigate further. Moreover, his inferences from the statistics would seem naive at best to one experienced in either the regulatory or the regulated community. It is these perceptions, however, that will attain a reality that will persist in the minds of any who read that paper and, for one reason or another, go no further. Only the individual with experience in the actual situations giving rise to apparent bias is in a position to put the statistical results into the perspective demanded in our multifaceted society. Ignorance allows the perceptions to assume a reality they may not deserve.

It is unfortunate that education is in a sense a goal of this article. Yeager was seeking to clarify the limits of law as a regulatory mechanism and to shed light on how regulation reproduces existing social orders. Whether he actually did so or whether he instead reinforced preexisting misconceptions cannot be determined from the article alone. Additional research by an insightful observer is required.[23]

It is one thing for academicians, who ordinarily remain aloof from ongoing policy formulation, to lend credence to questionable perceptions. It is quite another thing when the electorate takes political action on such perceptions. That is arguably the situation in California where voters can actually legislate by referendum.

In the widely reported Proposition 65 of the mid-eighties, the California electorate found, notwithstanding the major federal commitment to controlling toxic and hazardous substances, "that hazardous chemicals pose a serious potential threat to their health and well-being, that state governmental agencies have failed to provide them with adequate protection."[24] Significantly enough, Californians sought to be informed about exposures that could cause significant physiological harm but declined to set any priorities or to take into account quantitative considerations,[25] despite imposition of strict liability and draconian sanctions.[26] Here, public response to the technical and politicolegal difficulties associated with chemical substances control was simply to bypass both. Proposition 65 was based more in hysteria than in any relevant reality. As such, it was still making headlines in 1989.[27] One scientist, very much involved in politicolegal applications of his work, called Proposition 65 "a thoroughly silly law, with enormous cost and no gain in public health." Furthermore, it is a law that demands proofs that current science and technology cannot provide.[28] In brief, the mission is a noble one, but it has lost sight of everything but certain perceptions surrounding, in particular, carcinogens. Proposition 65 seems unlikely to advance the cause of scientific literacy or of decisions based in sound application of interdisciplinary principles.

Unfortunately, very little in this area is written for the lay reader, except to advance one private agenda or another. Writings for nonspecialists spend more time advocating a position in the strongest possible terms than in educating the reader with regard to the underlying sciences and other relevant disciplines.[29] There are, as one observer noted, too many who utter dollar opinions with a basis in a nickel's worth of information.[30]

It is all too easy for the reader, even when experienced in the area, to be caught up in the rhetoric. The uninitiated may never be able to escape the pervasive influence of propaganda to formulate an independent position with some basis in the foundational disciplines.

What is clearly before us as we enter the 1990s is whether our public policy is going to continue to respond to public demands or instead will go about the business of meeting public needs. We have much to accomplish before that basic alteration of course is a realistic prospect. We first have to reenfranchise society by overcoming scientific illiteracy in hope of fostering articulation of the real needs. Thus may public concerns be channelled in a more constructive direction,[31] for the full complement of roles in policy formulation.

REENFRANCHISEMENT

> "We need to inspire young people with the beauty of science and give them the confidence to understand and contribute to an increasingly scientific and technological society."[32]

As we explore the means of reenfranchising American society through literacy, particularly in the sciences and technology, it is instructive to reiterate a characteristic of science. Scientific competence is based not so much on instant recall of facts and theory but on "an ability to reason and to investigate natural phenomena systematically . . . [S]cience is not a body of facts so much as it is a way of knowing." It is based less on memorization of old knowledge and more on discovering new.[33]

In contrast, pupils in our schools have a rudimentary acquaintance with scientific facts and principles. At the age of seventeen, however, only 7 percent have the capacity to infer relationships or draw conclusions from detailed scientific knowledge. We must teach thinking skills. The question is whether we can correct our course.[34] Hope can not be particularly strong where it can be observed that few American adults know that electrons are smaller than atoms and that nearly half of this population believes early humans lived in the time of dinosaurs.[35]

The remainder of this chapter briefly addresses reenfranchisement. Starting with available tools, we will pause briefly to acknowledge inherent impediments to reenfranchisement in traditional approaches to education, broadly defined. Then we will acknowledge innovative approaches that are taking the initial steps on the journey to reenfranchisement. Finally, we will consider what the future may hold.

In order to tailor this admittedly exhaustive area to the thesis of this book, we will with few exceptions sample literature that is—or should be—popular, that is, available and accessible to the lay reader. Because discussion here is greatly reduced from the vast contemporary canon addressing national educational needs, a hierarchy of approaches will be observed in revealing existing tools and associated impediments to scientific literacy. That hierarchy will then be reversed as innovative approaches are sampled.

By way of guideposts, the hierarchy is comprised of institutions, more specific curricula and individual courses, and less formal approaches, including the frankly experimental and perhaps prophetic. In keeping with national institutions, there is a subdivision among educational programs, politicolegal approaches, and approaches emerging in the marketplace.

The Tools

> "Realizing that most citizens do not understand the technology (including the relevant science) that they are in daily contact with, and on which they vote, is the key."[36]

Scientific literacy has been defined by a coalition of scientists, teachers, school administrators, and policy analysts as the ability to read with comprehension news items on science and to be able to apply scientific information to personal decisions on emerging policy with a basis in science. Similarly, the individual should be able to engage in a scientifically informed discussion of a contemporary issue that depends on scientific facts and projections.[37] Furthermore, given the current direction of industry, employment increasingly demands fluency in some technical discipline.[38]

There seems to be little doubt about the need for scientific literacy. Neither is there much doubt about the existence of tools for meeting that need. The obvious starting point for the individual is the public schools.[39] The basic skills must be taught to all, starting early in formal education.[40] Acquiescence by default to an elite is both self-defeating and dangerous,[41] as well as completely avoidable.

We *are* capable of teaching excellent science to every pupil in the primary grades.[42] Our other institutions can assist.[43] As Robert M. Hazen has observed: "Science can provide an exhilarating outlet for every child's curiosity."[44] Education proceeds into undergraduate programs, which can play more than one major role in enhancing scientific literacy, as will be addressed below. The federal government also has major roles to play, not the least of which is the funding of innovative programs.[45] The marketplace is also very much involved, as has already been noted. As early as the mid-seventies, the American Bar Association had instituted a section of science and technology, which publishes *Jurimetrics Journal.*[46]

The tools are there. The problem is that traditions within the tools are frequently the impediments to their redirection for advancing a broadly based scientific literacy.

The Impediments

> "Technologists have not had the courage to tackle obsolete social structures, and have not been able to communicate to their fellow citizens how technology can . . . empower them and lead them to better health care, education and municipal services."[47]

Popular author Louis L'Amour left school at fifteen. Along with economic necessity, L'Amour asserted as a reason that public school was interfering with his education. His argument was with our approach to education, which he castigated as pedestrian and inflexible. The result is retardation of children's innate eagerness to expand their horizons and a stunting of their innate logic. In sum, L'Amour charged that we educate people not to think but to hold opinions.[48] Nowhere are such failings more intensive than in the matter of scientific literacy. Teaching in the public schools has been held out as the most serious problem.[49] We still promote opinion more than we do thinking, and today the latter is needed more than ever before.

There are those who assert that we are hardly teaching science in the elementary grades, if at all.[50] More than one young student is turned aside from science by pedantic methods. Through rote and mindless memorization, youngsters find science boring, difficult, and irrelevant.[51] Scientists themselves must accept a major share of the blame. It helps not at all when a student who may well want to discover science is forced to perform a dreary exercise at best distasteful, at worst a violation of a nascent personal ethic—a student, for example, who believes with a certain justification that she can learn biology without being required to dissect a frog and who has the intellectual capacity to suggest an alternative her teachers and others with a putative interest in scientific literacy summarily dismiss.[52] There is, in short, an unfortunate distinction between what scientists, teachers, school administrators, and policy analysts believe to be important and what is actually being taught.[53]

The whole problem cannot be attributed to teachers. In addition to prevailing institutional attitudes, there is the matter of how we test youngsters. When we test for recall of facts and theory at the expense of reasoning abilities, that is what teachers will instill.[54]

What is occurring in primary and secondary education depends on what is instituted in our undergraduate programs. There are some unfortunate perceptions there. Not all relate directly to scientific literacy, although so-called liberal education is the context for science as presented in the public schools.

On one level, there is the matter of what undergraduates and their parents are perceived as demanding. There are those who believe that they are seeking courses which lead directly to lucrative careers. Within a faculty, there is the problem of protecting turf. Most scientific faculty want no part of "baby biology" and its counterparts in other sciences. Then there is the matter of the perceptions from high atop that ivory tower. Salient among these is a profound distrust of chemicals—"poisons"—and the industry that produces them along with profits and lies. Such perceptions lead to a yearning to halt technological progress and to turn back the clock on technology.[55]

Unfortunately, teaching undergraduates for the purpose of expanding scientific literacy among the populace is not considered the best path to success among fellow faculty.[56] For whatever reason, prospective scientists are far more likely to venture into the humanities than the reverse; humanities majors are not made welcome in science departments. In 1987 it could be asserted that our best schools foster continuation of C. P. Snow's two cultures.[57] The strict maintenance of boundaries isolating academic departments is a major impediment to broadening educational horizons among undergraduates and those they will subsequently influence.[58] But there are few, if any, suitable texts.[59]

For the most part, it seems that any effort to overcome the strictures imposed by educational institutions are purely ad hoc and subject to the commitment and imagination of individual educators.[60]

Moving into the politicolegal system, we find institutional impediments prevailing as recently as 1991—and at the EPA, which has enjoyed sufficient experience to have learned better. That agency's panel reviewing potential health threats attributable to high-power transmission lines had neglected to include a physicist or even a biophysicist.[61] Only three years earlier representatives of the agency had noted: "The compartmentalization of environmental problems has created a difficult climate for analyzing or resolving complex multimedia issues."[62] Thus compartmentalization and bureaucratization would appear to be ubiquitous among our institutions. Moreover, there are major discrepancies among sectors of public and private America with regard to the nature of and proper agenda for environmental policy formulation,[63] as one prime example of disjuncture between societal needs and societal preparation to meet them.[64]

As far as the marketplace goes, science and technology have a partner for enhancing the education of the public in the legal profession, but that profession has internal and external problems of its own. Scholarly writing in law is as obtuse as writing can become. This situation has been attributed to a professional inferiority complex and the writings characterized as lacking serious effort or rigor.[65]

Superimposed on any internal lapse in communicative skills is the rapidly expanding gap between science and a "stumbling, misinformed legal system" where in education of the judiciary, for one, is rendered impracticable by sharp disagreements among scientists who venture into the courts.[66] Then there is the call for demystification of the legal system and its jargon on behalf of the public at large.[67]

That entity charged with observation of all sectors of society, the press, does not escape castigation. Instead, failures among the professions to communicate internally or to outsiders is compounded by "the propensity of the press to make sweeping generalizations and the frequent inability of journalists to go beyond superficialities in reporting science related to complex regulatory issues."[68] Unfortunately, news reports, no matter how emotional, speculative, misleading, and simply wrong, have a ring of authenticity not readily dissipated.[69] Still, we are not prepared to concede defeat.

Prospective Remedies

> "We have tried money without imagination, and it has failed. We have provided imagination without money, and it has failed. Perhaps . . . we can provide both ingredients simultaneously. It might be fair, effective, and fun all at the same time."[70]

The environmental movement is proving to be just the impetus we need. In 1988 a resurgence in interest in environmental issues was reported, as was strong public demand for more information. Schools were demanding materials to assist in teaching about the environment.[71] The interest has now carried beyond environmental issues. In recent years there have been responses to the broader challenges of scientific illiteracy throughout our institutions.

In the federal government the response has taken a variety of forms, including a call for a "regulatory budget," requiring allocations to be integrated for the setting of priorities.[72] Within existing institutions, the National Science Foundation budgeted 35 percent of its education dollars for elementary school programs.[73] In 1989 the research vice president for Argonne National Laboratory called for a multilateral national commitment to science education.[74] At the beginning of 1990 nearly twenty-five bills promoting science education were pending in Congress.[75] Elsewhere businesses were urged to provide tutors in math and the sciences,[76] and "techies" were being trained to communicate with marketers, accountants, and their own colleagues for effective promotion of individual research and development efforts.[77]

Among scientific institutions there were such programs as the Westinghouse Award for Public Understanding of Science, cosponsored by the American Association for the Advancement of Science (AAAS), the fifteenth annual AAAS colloquium on science and technology policy, addressing science and technology and the changing world order, and three ambitious national projects to "save" science education, notably that of AAAS, entitled Project 2061:Science for All Americans.[78]

Legal education has also taken up the gauntlet. The president of the Association of American Law Schools called for building bridges, with express reference to those between and among diversified intellectual perspectives, including economics and statistical analysis, and between law schools and other professional schools.[79] Another lawyer hailed the advent of an innovative legal text that interweaves with traditional materials such novelties as the sociology of administrative agencies.[80] One legal scholar admitted to investigating how legal processes respond to changes wrought by scientific advances and even suggested that law students be exposed to sufficient technical materials to familiarize them with scientific methods. The avowed goal was not to produce a new breed of lawyer-scientist or scientist-lawyer, but for each discipline to achieve common goals.[81] The American Bar Association included among nine goals that of increasing the public's understanding of law and the legal profession.[82]

The press, in cooperation with other institutions, has also been making strides.[83] Interdisciplinary and advanced liberal art degrees in the sciences, often to prepare teachers for the public schools, are proliferating.[84] Sometimes a particularly imaginative educator goes far in advancing scientific literacy among those who will never become scientists.[85] Some of this kind of effort eventually can be felt in the primary and secondary grades.[86]

Valuable programs in public education are expanded and the failures of formal education can be partially remedied through programs that are experimental, prophetic, or merely ad hoc. Established programs dedicated exclusively to the kind of expansion promoted here include the National Association for Science, Technology, and Society at one level and the American Scientific Affiliation at another. Narrowly focused but no less valuable is the National Organization for Rare Diseases.

There are more modest efforts, equally significant in their own way: AAAS has sponsored volunteer programs including Scientists and Schools at Chicago's Museum of Science and Industry. The Vermont Institute of Natural Science has been selected as one of sixty centers nationwide in the Alliance for Environmental Education's National Network for Environmental Education, a community outreach effort. TOPS Learning Systems out of Oregon publishes *Ideas* to promote hands-on learning in the classroom. Vermont's Keewaydin Environmental Education Center has a resident school program. These are but a few examples of activities around the nation.

Innovative approaches throughout our institutions are frequently announced in a variety of publications. Examples are a kind of ROTC for engineers,[87] peer reviews of medical testimony,[88] and judicial impact statements analogous to Environmental Impact Statements pursuant to NEPA.[89]

But in a sense there is nothing new in all these efforts, except perhaps a pervasive sense of urgency. Individuals have been contributing to the cause of broad scientific literacy all along. The creator of Peter Rabbit is one example.[90] A journal called simply *Risk* is a more recent, more focused example for professionals.[91] Individual scientists are encouraged to attend meetings and hearings, to take time to talk with the media, and to deal with legislators.[92] Teachers[93] and professors[94] invent unique approaches appealing to the imagination. Students make commitments to communication across professional lines.[95]

What, then, is still lacking? In addition to breaking down persistent institutional barriers and reaching an attentive public, what we have yet to institute is a multidisciplinary, truly integrated approach to the issues before us. Broadly defined, federal environmental protection and pollution-control policy formulation is the epitome of the needs and at least suggests some of the answers.

WHERE ARE WE GOING?

> "Our children should be given the chance to explore backward in time, look outward through space and discover unity in the workings of the cosmos."[96]

Our politicolegal institutions demand that all voices be heard if not necessarily heeded. Although it is unlikely this country will ever speak to any issue with a single voice, we can and must ensure that less alteration occurs between information and response. Warping of information and simple myopia are particular risks to us in areas at the intersection of law, science, technology, and policy, as has been illustrated here with specific reference to pharmaceutical innovation and the environmental movement in the context of reactions to innovative technologies. Attendant difficulties arise at a personal level and, ultimately, on a global scale.

The multitude of voices *can* be provided the opportunity to rise out of a foundation of scientific literacy rather than one of ignorance and imaginative speculation shading into hysteria. That foundation in turn can be better coordinated and holistic. The ideal of coordinated regulatory approaches based in sound reasoning from fact can be brought closer to the real world. We have the interest and an active core around which innovative programs can be built.

That core exists throughout our institutions, but it is far from consolidated, and it is encouraged mostly from outside each institution. There are, moreover, major impediments to needed change, too often located within each institution and in unrealistic demands from without. The more we understand our institutions and the factual bases for their decisions, the better we can effect public needs while accommodating reasonable demands from the full array of affected sectors. One model for the problems is found in the

milieu of regulation accompanying pharmaceutical innovation. Lessons guiding a holistic response are beginning to emerge in environmental planning, with special reference to environmental control of products of recombinant DNA.

Too often the lessons are imparted more by default than by design. We must take a more active approach to resolution through the interdisciplinary, multilateral integration that includes a literate electorate. The cynics who sneer and the idealists who smile superior smiles—and turn away, defeated—must come closer together as well.

NOTES

1. Albert Einstein, quoted in J. LILLY, COMMUNICATION BETWEEN MAN AND DOLPHIN:THE POSSIBILITIES OF TALKING WITH OTHER SPECIES (1987), at vii.

2. Adapted from a presentation by Jonathan Lash, Director, Environmental Law Center at Vermont Law School, workshop on Current Issues in the Environment, Environmental Conference in South Royalton, Vermont, Apr. 13, 1991. Mr. Lash observed that while he was in government he and his colleagues were working in an environment that was almost entirely data-free.

3. Einstein, *supra* note 1.

4. Westheimer, *Are Our Universities Rotten at the "Core?"*, 236 SCI. 1165, Jun. 5, 1987 Policy Forum, at 1166.

5. K. NICHOLS, TECHNOLOGY ON TRIAL:PUBLIC PARTICIPATION IN DECISION MAKING RELATED TO SCIENCE AND TECHNOLOGY (1979), at 109–111 *passim*.

6. To act on needs at the expense of demands is an act of profound political courage, perhaps beyond our present capacity. For all their significance to meaningful resolutions, courageous political decisions lie far outside the scope of this book.

7. Orr, *Ecological Literacy: Education for the Twenty-first Century*, Fall 1989 HOLISTIC EDUC. REV. 48, at 49. David W. Orr is the founder and codirector of the Meadowbrook Project, an ecological education center in Arkansas, and Associate Education Editor of *Conservation Biology*.

8. Bacon, *College Seniors Fail to Make Grade*, WALL ST. J., Oct. 9, 1989, at B1.

9. *See*, e.g., Grogan, *Engineering's Silent Crisis*, 247 SCI. 381, Jan. 26, 1990 Editorial; Bloch, *Scientific and Technological Literacy: The Need and the Challenge*, 6 BULL. SCI. TECH. SOC'Y 138 (1986); Carter, 238 SCI. 969, Nov. 13, 1987 review of GENETICS IN THE COURTS by Henry M. Butzel; Pool, *Who Will Do Science in the 1990's?*, 248 SCI. 433, Apr. 27, 1990 News & Commentary; Bloch, *Education and Human Resources at the National Science Foundation*, 249 SCI. 839, Aug. 24, 1990 Perspective.

William R. Grogan wrote as Dean of Undergraduate Studies, Worcester Polytechnic Institute. Formerly with IBM, Erich Bloch wrote as the Director of the National Science Foundation.

10. *See*, e.g., Selby, *Technological Literacy: A National Imperative and Benefit*, 6 BULL. SCI. TECH. SOC'Y 146 (1980); Tressel, *A Strategy for Improving Science*

Education, Apr. 8, 1988 presentation to the American Education Research Association; Abelson, *Science at the Four-Year and Master's Universities,* 239 SCI. 705, Feb. 12, 1988 Editorial; Davis, *Panel Bemoans U.S. Scientific Illiteracy, Prescribes More Ideas and Fewer Facts,* WALL ST. J., Feb. 24, 1989, at B4; Culotta, *Can Science Education Be Saved?,* 250 SCI. 1327, Dec. 7, 1990 News & Comment; Koshland, *The Dirty Air Act,* 249 SCI. 1481, Sep. 28, 1990 Editorial. *See also* the tabloid pull out on education, *The Knowledge Gap,* WALL ST. J., Feb. 9, 1990. The lead headline is "Smarter Jobs, Dumber Workers, Is that America's Future?"

Cecily Cannan Selby was an adjunct professor of science education at New York University. Philip A. Abelson was Deputy Editor of *Science.* Elizabeth Culotta was commenting as a science writer for the *Milwaukee Journal. See* note 36, *infra.*

11. Bloch (1986), *supra* note 9.

12. Carter, *supra* note 9.

13. Tressel, *supra* note 10.

14. Abelson, *supra* note 10.

15. Koshland, *supra* note 10.

16. Mohl, *Report to State on Environmental Pollution Late; Author Says It Just Keeps Growing,* BOSTON GLOBE, Feb. 1, 1988, at 1. The "author" was former Massachusetts state representative Roger Goyette, Sr., authorized by the Massachusetts House Speaker to write a report on pollution.

17. UNFINISHED BUSINESS IN NEW ENGLAND: A COMPARATIVE ASSESSMENT OF ENVIRONMENTAL PROBLEMS (EPA/901/7–89–001 1988). It is this series of reports that led to the convening of the Science Advisory Board, whose conclusions are addressed in the preceding chapter.

18. Orr, *supra* note 7, at 48.

19. Koshland, *supra* note 10.

20. Henry Brehm, personal communication at Boston University's Center for Technology and Policy, October 21, 1987.

21. Goslant, *Citizen Communication and Administrative Discretion in the Cleanup of Narragansett Bay,* 12 HARV. ENVTL. L. REV. (2) 521 (1982), *passim.*

22. Yeager, *Structural Bias in Regulatory Law Enforcement: The Case of the U.S. Environmental Protection Agency,* 34 SOC. PROBS. (4) 330 (1987).

23. *See,* e.g., S. JASANOFF, THE FIFTH BRANCH:SCIENCE ADVISERS AS POLICYMAKERS (1990), and M. LANDY, M. ROBERTS and S. THOMAS, THE ENVIRONMENTAL PROTECTION AGENCY:ASKING THE WRONG QUESTIONS (1990).

24. Safe Drinking Water and Toxic Enforcement Act of 1986, Section 1.

25. *See* ibid., Sections 1(c), (d) and Chapter 6.6, Sections 25249.6, 25249.7(d), 25249.8(b), 25249.11 (f)

26. *See* Chapter 6.6, *passim* and Section 25249.7.

27. Anonymous, *A Corrosive Fight over California's Toxics Law,* 243 SCI. 306, Jan. 20, 1989 News & Comment.

28. Ibid. The scientist was Bruce Ames, a biochemist at the University of California.

29. For representative diatribes—at opposite extremes—*see* D. RAY, TRASHING THE PLANET:HOW SCIENCE CAN HELP DEAL WITH ACID RAIN, DEPLETION OF THE OZONE, AND NUCLEAR WASTE (AMONG OTHER THINGS) (with L. Guzzo 1990), and J. LASH, A SEASON OF SPOILS:THE

STORY OF THE REAGAN ADMINISTRATION'S ATTACK ON THE ENVIRONMENT (with K. Gillman and D. Sheridan 1984). For pertinent commentary including ethics considerations, *see* Jones, *Coping with Controversy: Conflict, Censorship and Freedom within Evangelicalism,* 40 PERSP. SCI. & CHRISTIAN FAITH 32 (1988).

30. Witbeck, *Is Fiberglass a Health Hazard?,* WALL ST. J., Mar. 19, 1988 letter to editor, at 27. Paul C. Witbeck is a certified industrial hygienist.

31. Lewis, *Hazardous Waste Facility Siting: Ultimate Authority Vested in Local Opinion,* 2 NAT. RESOURCES & ENV'T (2) 24 (1986). For the broader perspective, *see* Nichols, *supra* note 5, and Bugliarello, *Empowering Citizens through Technology Literacy,* 10 BULL. SCI. TECHN. SOC'Y 187 (1990).

32. Presidential candidate Dukakis, *Science Policy,* 242 SCI. 173, Oct. 14, 1988 Policy Forum, at 176.

33. Hartwig, *Better Testing Key to Better Scientific Learning,* WALL ST., J., Feb. 22, 1989, at A16. Mark D. Hartwig was on the National Science Foundation's committee addressing science testing.

34. Ibid.

35. Norman, *Science Indicators: Healthy . . . for Now,* 247 SCI, 803, Feb. 16, 1990 News & Comment. (Ellipses in original title).

36. Roy, *"Citizen-Friendly" Science,* 251 SCI., Jan. 25, 91 letter to editor (unpaginated) in response to Culotta, *supra* note 10. Rustum Roy, in the Materials Research Laboratory at Pennsylvania State University, is Editor in Chief of *The Bulletin of Science, Technology & Society,* published by the National Association for Science, Technology and Society. This journal routinely publishes relevant educational materials ranging from single courses to whole curricula.

37. *Scientific Illiteracy,* 246 SCI. 214, Oct. 13, 1989 Briefings.

38. *See,* e.g., E. USEEM, LOW TECH EDUCATION IN A HIGH TECH WORLD (1986).

39. *See,* e.g., Selby and Culotta, *supra* note 10; Bloch (1986), *supra* note 9; Koshland, *Scientific Competency through Fun,* 243 SCI. 989, Feb. 24, 1989 Editorial; Langreth, *School Science Surveyed,* 251 SCI. 1023, Mar. 1, 1991 News & Comment; Putka, *Learning Curve:Lacking Good Results, Corporations Rethink Aid to Schools,* WALL ST. J., Jun. 27, 1989, at 1; Hazen, *Why My Kids Hate Science,* NEWSWEEK, Feb. 25, 1991, at 7.

40. *See* Selby, *supra* note 10.

41. Bloch (1986), *supra* note 9.

42. Koshland (1989), *supra* note 39.

43. *See,* e.g., Putka, *supra* note 39, and Useem, *supra* note 38.

44. Hazen, *supra* note 35.

45. *See,* e.g., SUMMARY OF GRANTS, FY 1984–86 (Directorate for Science and Engineering Education of the National Science Foundation, (1987). NSF had been promoting "Public Understanding of Science" for twenty-five years at the time the summary was prepared.

46. *Proceedings of ABA Section of Science & Technology,* 19 JURIMETRICS J. (5) (1979).

47. Bugliarello, *supra* note 31, at 189.

48. L. L'AMOUR, EDUCATION OF A WANDERING MAN (1989), at 74–75.

49. Useem, *supra* note 38.

50. *See,* e.g., Davis, *Giving Students a Head Start in Science,* WALL ST. J., Dec. 24, 1990, at 9.

51. Culotta (1990), *supra* note 10; Hazen, *supra* note 39.

52. *See* Koshland, "Homo Photosyntheticus," 239 SCI. 449, Jan. 29, 1988 Editorial.

53. "Scientific Illiteracy," *supra* note 37.

54. Hartwig, *supra* note 33.

55. Based with minimal paraphrasing on conversations with a classicist and a historian on the faculty of a small, mid-Atlantic liberal arts college in the spring of 1988. Although the conversation admittedly is of limited perspective, the sentiments have been repeatedly confirmed from the time the author commenced graduate school in the early sixties into the present.

Published confirmations include that of a historian who has turned to historical novels as a teaching tool because graduate work was dry and too far removed from ordinary living. *See* Waite, *History for the Masses,* EAGLE TIMES (Claremont, N.H.), Apr. 22, 1991, at 1. *See also* Roy, *Brave New World,* January 1989 CONNECTICUT ALUMNUS, at 3.

That liberal education itself is besieged is suggested by items addressing withdrawal from traditional Western civilization courses. *See,* e.g., Barchas, *Stanford Would Toss Intellectual Heritage to the Winds,* WALL ST. J., Jan. 21, 1988, at 30; *The Stanford Mind,* WALL ST. J., Dec. 22, 1988 editorial, at A14. National political philosophy is one component of the divergence from tradition, at the expense of tradition, and is another component of unrest in our colleges and universities that will have considerable influence on how science is presented to the larger society. *See,* e.g., Rabinowitz, *Vive the Academic Resistance,* WALL ST. J., Nov. 13, 1990, at A22.

56. *See,* e.g., Palca, *Goring Sacred Cows for Education Reform,* 249 SCI. 1242, Sep. 14, 1990 Briefing, at 1243.

57. Westheimer, *supra* note 4, at 1165–1166 *passim.*

58. Boulding, *The Prospects of Economic Abundance,* in POLITICS AND ECOLOGY (P. Foss, ed., 1972), at 32–33.

59. Personal communication with a professor in the biological sciences at Boston University in 1988. Certainly, there is still no ecology text suitable for advanced-degree students who lack formal education in the sciences.

60. *See,* e.g., Pool, *Freshman Chemistry Was Never Like This,* 248 SCI. 157, Apr. 13, 1990 News & Comment.

61. *EPA: Physicists Unwelcome on EMF Panel,* 251 SCI. 863, Feb. 22, 1991 ScienceScope.

62. *Unfinished Business, supra* note 17, at iii.

63. *See* The Becker Institute Survey on the Agenda for the State and Relationships among Leadership Groups (January 1987). The state in question is Vermont.

64. Useem, *supra* note 38.

65. *See* Barrett, *To Read This Story in Full, Don't Forget to See the Footnotes,* WALL ST. J., May 10, 1988, at 1. (Or simply attempt to read any law review article.) That scientists *need* a partner is further suggested cross-disciplinary scientific illiteracy in their own ranks. *See* Pool, *Scientific Literacy: The Enemy Is Us,* 252 SCI. 266, Jan. 18, 1991 News & Comment.

66. Attributed to Henry M. Butzel, author of *Genetics in the Courts* (1987) in a review by Stephen L. Carter of Yale Law School, 238 SCI. 969, Nov. 13, 1987.

67. *Learning about Law on the Cheap,* WALL ST. J., Oct. 13, 1988 Law column, at B1.
68. Goldstein, *Science, Policy, and the Press,* 249 SCI. 111, Jul. 13, 1990 letter in response to Pool, *supra* note 65.
69. Carley, *In a Big Plane Crash, Cause Probably Isn't Everybody's Guess,* WALL ST. J., Mar. 29, 1988, at 1.
70. Koshland (1989), *supra* note 39.
71. *Annual Issue,* ERIC/CLEARINGHOUSE FOR SCIENCE, MATHEMATICS AND ENVIRONMENTAL EDUCATION (Ohio State University 1988), first page (unpaginated).
72. R. LITAN AND W. NORDHAUS, REFORMING FEDERAL REGULATION (1983), at 134.
73. *New Science Programs Try to Spark the Interest of Younger Students,* WALL ST. J., Jan. 11, 1990, at 1.
74. Massey, *Science Education in the United States: What the Scientific Community Can Do,* 245 SCI. 915, Sep. 1, 1989 Association Affairs.
75. *Spotlight on Education,* 247 SCI. 405, Jan. 26, 1990 News & Comment, at 405.
76. KIPLINGER WASHINGTON LETTER, Dec. 9, 1988.
77. *Making "Techies" Enter the Real World,* WALL ST. J., May 17, 1989, at B1.
78. *See* Holden, *Radical Reform for Science Education,* 243 SCI. 1133, Mar. 3, 1989 News & Comment, and Culotta (1990), *supra* note 10. AAAS has published status reports: PROJECT 2061:SCIENCE FOR ALL AMERICANS (1989) and THE LIBERAL ART OF SCIENCE:THE REPORT OF THE PROJECT ON LIBERAL EDUCATION AND THE SCIENCES (1990).
79. February 1991 AALS NEWSL., 1, at 2 and 3.
80. Brunet, *A New Breed of Law Book?,* 1985 LAW & ADMIN. 523.
81. Green, *The Law-Science Interface in Public Policy Decisionmaking,* 51 OHIO ST. L.J. (2) 375 (1990).
82. Gross, *Educating the Public,* September 1988 A.B.A. J., at 121.
83. *See* Baker, *C-SPAN: A Guide for Law Professors,* 40 J. LEG. PROF. (3) 295 (1990); Nicholson, *Communicating Science,* 249 SCI. 721, Aug. 17, 1990 (reporting on a AAAS program known as Mass Media Fellows, placing graduate scientists in journalism positions); *New Group Aids Enviro-Reporters,* 2 ENV'T TODAY (2) 12 (1991).
84. For example: Pennsylvania State University's Science, Technology and Society Program; the Graduate School of Public Affairs at the University of Washington; the Tufts Center for Environmental Management; Science, Technology and Environmental Policy at the University of Minnesota; Idaho State's Doctor of Arts in Biological Sciences (preparation for college teaching); the Union Institute's doctorate in Environmental Studies; and the Master of Studies in (Environmental) Law at Vermont Law School's Environmental Law Center.
85. *See* Pool, *supra* note 60, and Hazen, *supra* note 39.
86. *Demolishing the Layer Cake,* 250 SCI. 205, Oct. 12, 1990 Briefings; Davis, *supra* note 50.
87. Grogan, *supra* note 9.
88. Hays, *Doctors Facing Peer Reviews of Testimony,* WALL ST. J., Jan. 11, 1988, at 29.
89. Koshland, *Judicial Impact Statements,* 239 SCI. 1225, Mar. 11, 1988 Editorial.

90. Carpenter, *A Love of Nature—and Peter Rabbit Thrives at the Thornton Burgess Society*, BOSTON SUNDAY GLOBE, Jul. 31, 1988, at 50.

91. Inaugural issue published winter of 1990 by Thomas G. Field, Jr., at Franklin Pierce Law Center.

92. S. OLSON, SHAPING THE FUTURE:BIOLOGY AND HUMAN VALUES (1989), at 31.

93. Davis, *Teacher at Work:How Gifted Educator Does His Bit to Mold Tomorrow's Scientists*, WALL ST. J., Feb. 22, 1990, at 1.

94. Personal communication with F.H. Westheimer of Harvard, 1988.

95. *Student Profile*, U.S. NEWS & WORLD RPT., Oct. 15, 1990, at 120.

96. Hazen, *supra* note 35. Robert M. Hazen, a scientist, is coauthor of *Science Matters: Achieving Scientific Literacy*. *See also* Pool, *Science Literacy: The Enemy Is Us*, 251 SCI. 266, Jan. 18, 1991. *But see also* Culotta, *Science's 20 Greatest Hits Take Their Lumps*, 251 SCI. 1308, Mar. 15, 1991 News & Comment, challenging Hazen and his coauthor to take care in their use of language and to place *process* before facts.

EPILOGUE

"Future History"

As we set our sights on the twenty-first century, there is bright promise for the future on all fronts from individual health and welfare to world politics to global environments. The scientific endeavor can carry us far and need not pose untoward threats to humanity.

The optimist sees only the promise, the pessimist only the hazards. Both are real. It is, then, the realists among us who must navigate for the future. The promises of science and technology must be tempered to ensure not only wide distribution of benefits but also equitable distribution of costs. For us to steer a sound course, those benefits and costs must first be recognized and institutional responses crafted on the basis of comprehension. These needs demand societal direction throughout our marketplaces and within the politicolegal system. In the end, literacy, most especially scientific literacy, is the prerequisite.

It is easy to stand on the sidelines and offer solutions for someone else to turn into practice. It is quite another matter to hunker down and bring about needed changes in firmly established institutional traditions. Intentionally altering weather patterns may actually prove to be the less-ambitious undertaking. In its own way human nature is every bit as intractable as climatic factors—and about as easily comprehended.

Nevertheless, a protocol for enhancing merger at the intersection linking science to the humanities and to humanity, even in its altruistic sense, can be proffered. In doing so, the assumption is that existing institutions will gradually alter without being eliminated, through noncoercive multilateral pressures. The mission is not so much the formulation of different policy as a sounder process for rendering decisions in the name of the public. The goal, then, is to inform the interests—vested, parochial, societal—even though any merger is unlikely in the foreseeable future.

The protocol consists of five steps in approximate chronological order, with the first through fourth in pairs and the last a realistic goal for the foreseeable future: (1) institutional encouragement of effective communication between professions and society and (2) institutional encouragement of effective cross-disciplinary communication among professionals; (3) revision of undergraduate curricula and associated faculty administrative practices to educate non–science majors—initially for scientific literacy as defined here and then to teach scientific basics in primary and secondary schools—and (4) societal encouragement of news media to assume greater responsibility in conveying information, with editorializing clearly distinguished from reporting; and (5) establishment of a national program of scientific literacy (a) commencing in the primary grades, (b) incorporating an environmental component on the assumptions that ecology not only includes all the natural sciences but also in the foundation of the outstanding crisis of this era, the call to environmental stewardship, and (c) increasing attention to implications for sound policy formulation as students gain maturity.

The optimist may assume literacy can be instilled at will, while the pessimist might hold out little hope for it at all. Even the realist must be daunted by the enormity of the task. It will not happen in an instant. A scientifically literate electorate will only be established if we have the will and take the time. The process will almost certainly be more evolutionary than revolutionary, but that evolution is already in process.

Like all evolutions this one is marked with scattered alterations so minute as to be barely discernible, but the process is one of gradual building. Representatives of all our public and private institutions are perceiving the needs and taking individual action. Continuing the metaphor, the process is seemingly random. Some efforts will prove to be dead ends, but others will meet with success. In time the successes will gain momentum. In fact, they are already doing so.

American institutions move slowly. Deliberate change is inherent to our traditions, which have served us well throughout our national youth and coming of age. We cannot expect the merger of the whole mixture of perceptions and realities needed for sound policy formulation to occur overnight. It will take dedication, intensive effort and, above all, time. After all, such a consolidation demands a sea change in two salient institutions, education and journalism.

If they decline to step down from their ivory towers, academics must at the very least cross traditional disciplinary boundaries presently separating and balkanizing the sciences, both natural and social, and the humanities. The real change in direction will come with new generations, who depend, directly or indirectly, on higher education for their own literacy. Simultaneously, journalism must return to the news media, just as statesmanship must return to politics. These prospects must become more than fantasies of the hopelessly idealistic.

The prospect of literacy in the service of policy formulation at the intersections before us will never be met if the evolutionary process is terminated. The voices and efforts of one or a few here and there will be lost. But increasingly widespread expectations ultimately give rise to a groundswell of demand that must be met within our institutions if they are to endure.

At this juncture, even the optimist must acknowledge that the public at large has been traumatized by its own perceptions of science and technology. True advances have been spoiled by fear, mostly speculative, of the unknown and the misunderstood. Ethics and morality have been strained to the utmost as we seize newfound capabilities for intervening in natural processes for good or ill. Individually and hence politically, we are trapped in a transition stage characterized by denial, blame, fantasies of vengeance, or deadly resignation. We must work together to reach the level of productive reconciliation. To do so, we must begin to integrate the plethora of multidisciplinary individual efforts. We must demand true interdisciplinary communication throughout our institutions.

Numerous organizations are actually taking those first integrative steps. They represent scientific, politicolegal, social, and educational institutions. As a nation we are beginning to move from talking about relevant problems to doing something productive.

What may be needed to provide the impetus for an evolutionary leap forward is a forum where all the interested parties can encounter one another. An immediately arising query reflects the nature of the dilemma we face: What is the proper locus of such a forum? A chorus of nays will follow almost any designation in any sector, public or private. Hence, the forum itself must be independent, but interdisciplinary and integrated. Above all, it must be sparked with imagination.

If science can accept a theory called chaos, if individual scientists can convene an Asilomar, if EPA can recognize discontinuities in divergent goals espoused by mandate and the public at large, if AAAS and ABA can cooperate for the advancement of scientifically literate law, if academics and others can communicate in NASTS, if a NORD can respond to specific health needs with a balancing of individual and societal needs, if academic institutions can support an Environmental Law Center and comparable programs crossing traditional lines—surely there is more than hope for the future. There is also a viable germ of societal scientific literacy. The foregoing not only lead the way, they also form the corpus around which a novel forum can be established. Without abandoning its roots in existing institutions, this forum must transcend tradition in collecting and disseminating relevant information—as an initial step.

What is left to future history is to institute that forum, charge it with the necessary mandate to gather, consolidate, and publish promising approaches for a diverse readership and, finally, to lend the results effective credence. We have the means to meet the promises of the scientific endeavor in the next century. We must prove wrong those who assert we lack the will.

Selected Bibliography

GENERAL

Anonymous. *Magic and Medicine of Plants*. Pleasantville, NY : Reader's Digest, 1986.

Anteibi, E., and D. Fishlock. *Biotechnology: Strategies for Life*. Cambridge, MA: MIT Press, 1985.

Gough, M. *Dioxin, Agent Orange: The Facts*. New York: Plenum Press, 1986.

Lash, J., with K. Gillman and D. Sheridan. *A Season of Spoils: The Story of the Reagan Administration's Attack on the Environment*. New York: Pantheon Books, 1984.

Ray, D., with L. Guzzo. *Trashing the Planet: How Science Can Help Deal with Acid Rain, Depletion of the Ozone, and Nuclear Waste (Among Other Things)*. Lanham, MD: Regnery Gateway, 1990.

Rifkin, J. *Algeny*. New York: The Viking Press, 1983.

Snow, C. *The Two Cultures: And a Second Look*. 1963. Reprint. Cambridge, MA: Cambridge University Press, 1969.

Useem, E. *Low Tech Education in a High Tech World: Corporations and Classrooms in the New Information Society*. New York: The Free Press, 1986.

Worobec, M., and G. Ordway. *Toxic Substances Control Guide*. Washington, DC: The Bureau of National Affairs, 1989.

Zilinskas, R., and B. Zimmerman. *The Gene-Splicing Wars: Reflections on the Recombinant DNA Controversy*. New York: Macmillan, 1986.

Zimmerman, B. *Biofuture: Confronting the Genetic Era*. New York: Plenum Press, 1984.

POLITICAL SCIENCE

Litan, R., and W. Nordhaus. *Reforming Federal Regulation*. New Haven, CT: Yale University Press, 1983.

Lunch, W. *The Nationalization of American Politics*. Los Angeles: University of California Press, 1987.

Environmental Movement

Anonymous. *Selected Environmental Law Statutes*. St. Paul: West Publishing Co., 1990–91.
Caldwell, L. *Between Two Worlds: Science, the Environmental Movement, and Policy Choice*. Cambridge, MA: Cambridge University Press, 1990.
Davies, C., and B. Davies. *Politics of Pollution*. 2d ed. Indianapolis: Pegasus, 1975.
Hoban, T., and R. Brooks. *Green Justice: The Environment and the Courts*. Boulder, CO.: Westview Press, 1987.
Landy, M., M. Roberts, and S. Thomas. *The Environmental Protection Agency: Asking the Wrong Questions*. New York: Oxford University Press, 1990.
Magat, W., ed. *Reform of Environmental Regulation*. Cambridge, MA: Ballinger Publishing Co., 1982.
Tucker, W. *Progess and Privilege: America in the Age of Environmentalism*. Garden City, NY: Anchor Press, 1982.

THE NATURAL SCIENCES, BASIC AND APPLIED

Basic Science and Scientists

Abelson, P., ed. *Biotechnology & Biological Frontiers*. Washington, DC: The American Association for the Advancement of Science, 1984.
Graham, L. *Between Science and Values*. New York: Columbia University Press, 1981.
Koshland, D., ed. *Biotechnology: The Renewable Frontier*. Washington, DC: The American Association for the Advancement of Science, 1986.
Peters, J., ed. *Classic Papers in Genetics*. Englewood Cliffs, N.J.: Prentice-Hall, 1959.
Richter, M. *Science as a Cultural Process*. Cambridge, MA: Schenkman Publishing Co., 1972.
Sturtevant, A. *A History of Genetics*. New York: Harper & Row, 1965.

Science and Policy Formulation

Cherfas, J. *Man-made Life: An Overview of the Science, Technology and Commerce of Genetic Engineering*. New York: Pantheon Books, 1982.
Jasanoff, S. *The Fifth Branch: Science Advisors as Policymakers* Cambridge, MA: Harvard University Press, 1990.
Nichols, K. *Technology on Trial: Public Participation in Decision-making Related to Science and Technology*. Paris: Organization for Economic Cooperation and Development, 1979.
Thomas, W. ed. *Science and Law: An Essential Alliance*. Boulder, CO: Westview Press, 1983.

Pharmaceuticals and Pharmaceutical Regulation

Apfel, R., and S. Fisher. *To Do No Harm*. New Haven, CT: Yale University Press, 1984.

Asbury, C. *Orphan Drugs: Medical Versus Market Value.* Lexington, MA: Lexington Books, 1985.

Barber, B. *Drugs and Society.* New York: Russell Sage, 1967.

Davies, W. *The Pharmaceutical Industry: A Personal Study.* New York: Pergamon Press, 1967.

Grabowski, H. *Drug Regulation and Innovation: Empirical Evidence and Policy Options.* Washington, DC: American Enterprise Institute for Public Policy Research, 1976.

Mahoney, T. *The Merchants of Life: An Account of the American Pharmaceutical Industry.* Reprint. Freeport, NY: Books for Libraries Press, 1972.

Mintz, M. *By Prescription Only.* Boston: Beacon Press, 1967.

Statman, M. *Competition in the Pharmaceutical Industry: The Declining Profitability of Drug Innovation.* Washington, DC: American Enterprise Institute for Public Policy Research, 1983.

Talalay, P., ed. *Drugs in Our Society.* Baltimore: The Johns Hopkins Press, 1964.

Wardell, W., and L. Lasagna. *Regulation and Drug Development.* Washington, DC: American Enterprise Institute for Public Policy Research, 1975.

Index

About the Author

DOROTHY J. HOWELL is Adjunct Professor at the Environmental Law Center at Vermont Law School and a participant in environmental programs since 1969. She teaches a seminar in the multidisciplinary aspects of environmental planning, as well as courses in environmental sciences and toxics law.

www.ingramcontent.com/pod-product-compliance
Ingram Content Group UK Ltd.
Pitfield, Milton Keynes, MK11 3LW, UK
UKHW020818290926
15470UKWH00003B/1240

9 780899 306162